Stephan Rix

Laser Damage in Calcium Fluoride

Stephan Rix

Laser Damage in Calcium Fluoride
A theoretical approach to radiation damage

Südwestdeutscher Verlag für Hochschulschriften

Impressum/Imprint (nur für Deutschland/only for Germany)
Bibliografische Information der Deutschen Nationalbibliothek: Die Deutsche Nationalbibliothek verzeichnet diese Publikation in der Deutschen Nationalbibliografie; detaillierte bibliografische Daten sind im Internet über http://dnb.d-nb.de abrufbar.
Alle in diesem Buch genannten Marken und Produktnamen unterliegen warenzeichen-, marken- oder patentrechtlichem Schutz bzw. sind Warenzeichen oder eingetragene Warenzeichen der jeweiligen Inhaber. Die Wiedergabe von Marken, Produktnamen, Gebrauchsnamen, Handelsnamen, Warenbezeichnungen u.s.w. in diesem Werk berechtigt auch ohne besondere Kennzeichnung nicht zu der Annahme, dass solche Namen im Sinne der Warenzeichen- und Markenschutzgesetzgebung als frei zu betrachten wären und daher von jedermann benutzt werden dürften.

Verlag: Südwestdeutscher Verlag für Hochschulschriften GmbH & Co. KG
Heinrich-Böcking-Str. 6-8, 66121 Saarbrücken, Deutschland
Telefon +49 681 37 20 271-1, Telefax +49 681 37 20 271-0
Email: info@svh-verlag.de

Approved by: Mainz, Johannes Gutenberg-Universität, Diss., 2011

Herstellung in Deutschland:
Schaltungsdienst Lange o.H.G., Berlin
Books on Demand GmbH, Norderstedt
Reha GmbH, Saarbrücken
Amazon Distribution GmbH, Leipzig
ISBN: 978-3-8381-2810-8

Imprint (only for USA, GB)
Bibliographic information published by the Deutsche Nationalbibliothek: The Deutsche Nationalbibliothek lists this publication in the Deutsche Nationalbibliografie; detailed bibliographic data are available in the Internet at http://dnb.d-nb.de.
Any brand names and product names mentioned in this book are subject to trademark, brand or patent protection and are trademarks or registered trademarks of their respective holders. The use of brand names, product names, common names, trade names, product descriptions etc. even without a particular marking in this works is in no way to be construed to mean that such names may be regarded as unrestricted in respect of trademark and brand protection legislation and could thus be used by anyone.

Publisher: Südwestdeutscher Verlag für Hochschulschriften GmbH & Co. KG
Heinrich-Böcking-Str. 6-8, 66121 Saarbrücken, Germany
Phone +49 681 37 20 271-1, Fax +49 681 37 20 271-0
Email: info@svh-verlag.de

Printed in the U.S.A.
Printed in the U.K. by (see last page)
ISBN: 978-3-8381-2810-8

Copyright © 2011 by the author and Südwestdeutscher Verlag für Hochschulschriften GmbH & Co. KG and licensors
All rights reserved. Saarbrücken 2011

Für Heiner Opa

Contents

Acknowledgements		**vii**
1	**Introduction and Motivation**	**1**
2	**Calculation of material properties**	**7**
2.1	Describing Solid State Matter	7
2.2	Density Functional Theory	10
	2.2.1 The fundamental variables	11
	2.2.2 The Hohenberg-Kohn theorems	11
	2.2.3 The Kohn-Sham approach	12
	2.2.4 Exchange-correlation functionals	15
	2.2.5 Pseudopotentials	16
	2.2.6 Wave function expansion	17
2.3	Calculating diffusion properties	17
	2.3.1 Transition State Theory	18
	2.3.2 The climbing image nudged elastic band method	20
	2.3.3 Calculating phonon modes	22
2.4	Codes and Resources	24
3	**Properties of defect structures in CaF_2**	**26**
3.1	Pure CaF_2	26
3.2	Point defects in CaF_2	26
	3.2.1 The F center	26
	3.2.2 The M center	29
	3.2.3 The V_k center	29
	3.2.4 The H center	31
	3.2.5 The self-trapped exciton	31
3.3	Stabilized point defects in CaF_2	34
	3.3.1 Self-stabilization by M center formation	35
	3.3.2 Stabilization by impurities	35

	3.4	Diffusion properties of point defects	37
		3.4.1 F-center diffusion properties	37
		3.4.2 H-center diffusion properties	40
		3.4.3 Dielectric measurements on CaF_2	44
		3.4.4 Discussion of the results	45
	3.5	Optical properties of Ca colloids in CaF_2	48
		3.5.1 Lambert-Beer law	48
		3.5.2 Optical properties of CaF_2 and Ca	50
		3.5.3 Cross sections from Mie theory	51
	3.6	Formation energy of Ca colloids	54
		3.6.1 Binding energy of an F center to a colloid	55
		3.6.2 Energy of the Ca/CaF_2 interface	55
		3.6.3 Mechanical stress in colloids	56
		3.6.4 Formation energy	58
4	**Phenomenology of laser damage in CaF_2**		**60**
	4.1	Characterization of laser damage	61
	4.2	Rapid damage	62
	4.3	Long term laser damage	63
	4.4	AFM measurements	67
	4.5	Annealing by tempering	67
	4.6	Discussion of the observations	70
5	**A laser damage model**		**72**
	5.1	The pre-irradiation material	72
	5.2	Irradiation induced defect formation	72
	5.3	Defect agglomeration	74
	5.4	Diffusion based laser damage model	75
	5.5	Limits of stability of Ca colloids in CaF_2	78
	5.6	Reversibility of laser damage	78
	5.7	Preventing laser damage	79
6	**Conclusion**		**80**
A	**Cross sections from Mie-theory**		**84**
B	**Results for CaF_2 and relevant defects**		**91**
	Bibliography		**103**

List of Figures

2.1 Schematic illustration of the nudged elastic band method. 20

3.1 The fluorite structure. 27
3.2 Band structure and density of states of pure CaF_2. 27
3.3 The structure of the F center. 28
3.4 ELF and spin density of the F center. 28
3.5 Band structure and density of states of the F center. 29
3.6 The structure of the M center. 30
3.7 ELF of the M center. 30
3.8 Band structure and density of states of the M center. 31
3.9 The structure of the H center. 32
3.10 ELF and spin density of the H center. 32
3.11 Band structure and density of states of the H center. 33
3.12 The diffusion barrier of the F center 38
3.13 The diffusion path of the F center. 39
3.14 Images of the minimum energy path of the F center diffusion. 39
3.15 The diffusion path of the H center. 41
3.16 The diffusion barrier of the H center. 41
3.17 Images of the minimum energy path of the H center diffusion. 42
3.18 Conductivity of CaF_2. 44
3.19 The diffusion barrier of F_{Na} center formation. 47
3.20 Schematic illustration of extinction, absorption and scattering. 49
3.21 The refractive index of Ca. 52
3.22 Wavelength-dependent extinction, absorption, and scattering coefficients of Ca colloids in CaF_2 calculated from Mie-theory. 53
3.23 Colloid-size dependent extinction, absorption, and scattering coefficients and cross sections of Ca colloids in CaF_2 at 193 nm. 54
3.24 Surface energy of the Ca/CaF_2 interface for different Ca layer thickness. 57
3.25 Surface energy of the Ca/CaF_2 interface for the [100] and [111] orientation. . . . 57

3.26	Formation energy of a Ca colloid in CaF_2.	59
4.1	Induced extinction in CaF_2.	61
4.2	Extinction spectra of CaF_2 after irradiation with 10^4 pulses.	62
4.3	Induced extinction with fitted extinction coefficients.	64
4.4	Absorption and scattering in CaF_2.	66
4.5	NC-AFM measurements of a cleaved CaF_2 surface.	68
4.6	Annealing of CaF_2.	69
5.1	Diffusion model of laser damage.	74
5.2	Spatial concentration profile of F and H centers after 1 month.	76
5.3	Spatial concentration profiles of F and H centers at different temperatures and times.	77
B.1	Calculated structure, DOS, and band structure of CaF_2.	92
B.2	Calculated structure, DOS, and band structure of the F center.	93
B.3	Calculated structure, DOS, and band structure of the M center.	94
B.4	Calculated structure, DOS, and band structure of the H center.	95
B.5	Calculated structure, DOS, and band structure of the F_{Na} center.	96
B.6	Calculated structure, DOS, and band structure of the M_{Na} center.	97
B.7	Calculated structure, DOS, and band structure of a Na^+ impurity.	98
B.8	Calculated structure, DOS, and band structure of the H_Y center.	99
B.9	Calculated structure, DOS, and band structure of a Y^{3+} impurity.	100
B.10	Calculated structure, DOS, and band structure of the F_Y center.	101
B.11	Calculated structure, DOS, and band structure of the M_Y center.	102

List of Tables

3.1	Calculated formation energy of the F-H pair in CaF_2.	34
3.2	Stabilization energies of point defects.	35
3.3	The ionic radii of calcium, sodium, and yttrium.	35
3.4	Jump rate and prefactor for the diffusion coefficient of the F center.	40
3.5	Jump rate and prefactor for the diffusion coefficient of the H center.	43
3.6	Diffusion barriers of the F and the H center.	46
3.7	The Sellmeier coefficients for CaF_2.	51
3.8	Elastic stiffness and elastic compliance constants of Ca and CaF_2.	55
3.9	The lattice constants of Ca and CaF_2.	56
3.10	Formation energy of a Ca colloid in CaF_2.	58

Acknowledgments

This book is the outcome of my doctorate studies at the Johannes Gutenberg-Universität Mainz. My research was performed at Schott and Princeton University. Funding for my research was kindly provided by Schott and the Graduate School of Excellence *Materials Science in Mainz*.

Several people have contributed time and effort to this work and I would like to express my gratitude towards all of them. First and foremost, my supervisor Martin Letz made this unique opportunity possible and provided support, advice and guidance throughout this time. Ute Natura contributed experimental data and her profound experience in many discussions. During my time in Princeton, Ulrich Aschauer introduced me to QE and neb calculations. For the academic supervision, I thank Claudia Felser.

Further, I am grateful for all the support I received from my colleagues and co-workers at Schott, former Schott Lithotec and Hellma Materials, Prof. Felser and her group, Prof. Selloni and her group in Princeton, all MAINZ students, and anyone I may have forgotten.

Last but not least, I am indebted to my family for their support and encouragement.

Chapter 1

Introduction and Motivation

Calcium fluoride (CaF_2), also known as fluorite, is a crucial optical material for many modern applications. In 1888 Otto Schott discussed the introduction of fluorine into optical glasses with a low refractive index, because a high fluorine content would increase the glass's transmission in the blue and ultraviolet (UV) range while significantly lowering its dispersion [1]. Such a glass was desirable for the construction of an apochromatic lens, i.e. a lens system free of chromatic or spherical abberations, but was ruled out due to difficulties in manufacturing. In 1890 Ernst Abbe first mentioned the usage of fluorite for optical purposes [2], as it exhibits the unique properties Schott tried to accomplish in a glass. As early as 1884 Carl Zeiss, a collaborator of Schott and Abbe, had constructed the first microscope containing natural fluorite according to Abbe's calculations. Almost 130 years later synthetic CaF_2 is still the material of choice providing the same functionality for high-precision apochromatic optics in high-end cameras, such as high-definition television (HDTV) cameras [3, 4]. The early efforts by Schott and Abbe were the beginning of an important development, as CaF_2 and fluorine containing glasses are crucial components in many optical applications today [5].

Due to its high transparency ranging from 130 nm in the deep-ultraviolet (DUV) to 9 μm in the far infrared (IR) [6], CaF_2 is used for optics in the whole spectrum, but becomes irreplaceable at the extreme edges. In the IR range, it is used for optics in medical, military, and telecommunication applications. CaF_2 has also become an important material as a substrate for lab-on-a-chip applications, where it allows spectroscopic measurements at a wide range of wavelengths [7, 8].

However, the most demanding application for CaF_2 is DUV optical microlithography, where CaF_2 is a key lens material. Since commercial microchip production started in the early 1960s, the semiconductor industry has developed into one of the most important industries imaginable today. This development has been driven by the steady need for more computing power, which means faster and more efficient microchips. As early as 1965 Gordon E. Moore predicted that the number of transistors per unit area on commercially available chips would double every year [9]. He later corrected the doubling rate to about every 18 months, and this prediction, known as Moore's Law, holds until today.

Making microchips faster and more efficient essentially means shrinking their dimensions in order to control the heat dissipation on the chips, which in turn allows for higher clock rates. Currently microchips are produced with a critical dimension of 32 nm. The structuring of these microchips is done with optical lithography [10]. Since the critical dimension of optical lithography is directly proportional to the structuring wavelength, the application wavelength for optical microlithography has been reduced over the decades and is currently 193 nm, which is the wavelength of an ArF excimer laser. For some time the future of optical lithography seemed to be 157 nm-technology, where CaF_2 would have been the exclusive optical material. However, due to several technical issues the semiconductor industry decided to improve the resolution by immersion lithography and double-patterning techniques in use today. On-going research and development efforts aim at extreme-ultraviolet lithography (EUVL) as the future technology performing at 13.5 nm using reflective optics [11].

Because of the extremely high precision necessary for manufacturing structures in the range of 32 nm, the requirements for the optical systems and their components are extremely demanding. In addition to extremely high requirements on the homogeneity and the temperature-stability of the refractive index, an extremely high laser stability is required, i.e. the material needs to maintain stable optical properties under 193 nm-radiation at fluences of up to 1 J/cm^2 for several months. In order to achieve this an extreme purity of the material is necessary. In recent years the manufacturing of monocrystalline CaF_2 has improved to the extent that today CaF_2 can be produced with an impurity content below chemical measurability, i.e. below the parts per billion (ppb) range, and a negligible number of crystal defects. For microlithography only CaF_2 single crystals of highest quality, i.e. the highest purity and a minimal number of crystal defects are used.

At 193 nm only fused silica and some fluoride crystals can be considered as lens materials due to their transparency. Because of its cubic structure with nearly perfect optical anisotropy [12, 13], CaF_2 shows only very little birefringence. In addition, its chemical stability and its mechanical properties make it suitable for lens fabrication. Its extreme laser stability makes CaF_2 the key material for this application along with fused silica.

However, after irradiation with 193 nm for several months, CaF_2 of the highest quality still shows coloration effects. The investigation of these coloration effects, the so-called laser damage, has been studied intensively in recent years [14, 15, 16, 17] and is the topic of this work.

Historically, the coloration of fluorite (natural CaF_2) has been studied widely. In the early 1920s Cornelius Doelter investigated the visible coloration and decoloration of fluorite and many other minerals due to α, β, γ, and ultraviolet (UV) radiation and heating [18, 19, 20]. He suggested that the coloration effects he observed were due to the formation of colloidal particles in the minerals according to Mie's theory, which was published roughly ten years earlier [21]. His assumptions were confirmed by Luise Göbel [22] and Karl Przibram [23]. Przibram also showed that many coloration effects can be associated with impurity ions, especially with rare-earth ions [24, 25]. The coloration of natural fluorite and doped CaF_2 with a multitude of impurity ions in combination with color centers was extensively studied by Hans Bill and co-workers, who used spectroscopic measurement techniques such as electron spin resonance (ESR) [26, 27, 28, 29] and electron nuclear double resonance (ENDOR) [30, 31, 32].

In the 1970s color centers in fluoride crystals were studied extensively [33, 34]. In this time the focus shifted from understanding coloration of halide crystals to their application in new physical technologies. The color centers in halide crystals were thoroughly investigated with respect to their applicability in color center lasers [35]. These lasers were of special interest at the time as they promised to advance efforts to build tunable solid state lasers. In this context it was first shown, that laser interaction of doped CaF_2 showed optical damage related to point defects [36]. Additionally, impurity doped CaF_2 was also considered as a host material for optical data storage via persistent spectral hole burning on thin films [37].

Furthermore, CaF_2 has been studied as a model system for uranium oxide since the 1960s. Since both materials have the fluorite structure in common, it has been used to investigate diffusion properties [38], phononic properties [39], powder compaction properties [40], and thermal conductivity [41]. Apart from being a model system for uranium oxide, CaF_2 is also important as a lining material for crucibles in uranium production [40].

With the application of CaF_2 in microlithography starting in the late 1990s, the focus shifted to understanding excitonic processes caused by laser irradiation [42, 43, 44]. As CaF_2 is one of the crucial lens material for microlithography, the above mentioned research on understanding laser damage and the improvement of the optical performance of the material has been a major contribution to the performance of computers today.

The tremendous increase of computing power in the past decades has greatly benefited computational science. Nowadays, many material properties can be calculated due to the constant development and improvement of new algorithms, which rely on the availability of computing power. In conjunction with experiments, the theoretical prediction of material properties will become more and more important, because the theoretical approach allows for finding parameters and mechanisms which are experimentally not accessible. Additionally, the systematic sampling of many different materials or material configurations can be performed faster and cheaper with computational methods. Therefore, computational science is expected to make a substantial contribution to finding new functional materials in the future. While today's computational methods are mostly used to calculate the properties of a given material, the importance of the approach of reverse engineering, i.e. the determining of a material with the desired properties, will increase significantly in the future [45].

There are numerous theoretical methods for the calculation and prediction of material properties, which can be classified by various criteria [46]. Empirical force fields methods or classical molecular dynamics (MD) rely on Newtonian mechanics and fixed atom force fields. Macroscopic properties of a material are derived from the dynamics of the ionic movement, not taking into account the exact electronic structure. These methods allow the simulation of a large number of atoms in the range of one million, but have limited predictive capabilities. So-called *ab inito* methods on the other hand derive material properties based on the electronic structure, which is calculated with first-principle quantum mechanical methods, which do not rely on any empirical input. Since the quantum mechanical description of a material involves the solution of a many-electron problem, *ab initio* methods are limited to a much smaller number of atoms, in the range of a few hundred. Car-Parrinello molecular dynamics (CPMD) combines the classical molecular dynamics and a quantum mechanical approach [47], allowing for the simulation of the dynamical properties of a few hundred atoms limited to a timespan of a few picoseconds.

The *ab inito* approaches are faced with the difficulty of solving the Schrödinger equation for a many-electron system, of which in 1930 Dirac said it is "far too complicated to be practicable" [48]. The most common approaches to this solution today are the Hartree-Fock method [49, 50], density functional theory (DFT) based on the Hohenberg-Kohn theorems [51] and the Kohn-Sham approach [52], and quantum Monte Carlo (QMC) methods [53, 54]. The main difference between these methods is the way they incorporate electron exchange and correlation. While the Hartree-Fock method takes electron exchange into account, it neglects electron correlation. DFT is in principle an exact theory, that reduces the many-particle Schrödinger equation to the effective one particle Kohn-Sham equations [52]. However, electron exchange and correlation are described by an exchange-correlation functional, which is *a priori* unknown and therefore approximated. QMC on the other hand uses statistical integration methods to solve the multi-dimensional Schrödinger equation. In principle, QMC yields an exact solution of the Schrödinger equation. However, it is computationally very demanding and in practice limited to high temperatures.

The first approach to describing a quantum system with a density functional was made by Thomas [55] and Fermi [56, 57] in 1927, not taking into account any exchange or correlation effects. In 1930, Dirac extended this approach by formulating the approximation for local exchange [48]. 34 years later, Hohenberg and Kohn laid the foundation for modern DFT by formulating DFT as an exact theory of many-body systems [51]. By replacing the many-body problem with an independent-particle system and placing all exchange and correlation phenomena in a local exchange-correlation functional, Kohn and Sham provided an important step towards the solution of many-body problems [52]. In principle, the Kohn-Sham equations yield an exact solution of the many-body problem, which is only limited by the approximation of the exchange-correlation functional and the underlying assumptions such as the Born-Oppenheimer approximation.

The simplest approach to the Kohn-Sham exchange-correlation functional is the local density approximation (LDA), which assumes that the electron exchange and correlation of the many-electron system is the same as for the homogeneous electron gas. In spite of its simplicity, LDA has been remarkably successful, especially for systems with a fairly homogeneous electron density. However, it was the development of the general gradient approximations (GGAs) which increased the accuracy and established DFT as a widely accepted computational method. The most commonly used GGAs were proposed by Becke (B88) [58], Perdew and Wang (PW91) [59], and Perdew, Burke, and Enzerhof (PBE) [60].

For the description of transition metals and rare-earth ions with occupied $3d$ and $4f$ states, more sophisticated functionals are needed. One approach is to use orbital-dependent functionals such as LDA+U, which is based on LDA or GGA, but adds a Hubbard-like interaction to highly localized orbitals [61]. Another approach is to mix the Hartree-Fock exchange with density functional approximations, yielding hybrid functionals such as B3LYP, which combines Becke's exchange functional (B88) with the LYP correlation [62]. However, orbital-dependent and hybrid functionals both rely on parameters which need to be fitted to data.

In spite of its success DFT cannot answer all question regarding solid state materials and has some limitations. First, DFT is a ground state theory, which means that excited states are not necessarily described correctly. However, the prediction of excited states is quite accurate for many systems. Second, the correct description of the full electronic structure of $3d$ and $4f$-electron systems is difficult due to the strong electron correlation, since the explicit correlation is lost by reducing the many-particle wave function to effective one-particle wave functions. Third, the underlying Born-Oppenheimer approximation explicitly decouples the electronic system from the dynamics of the nuclei. Therefore, dynamic electron-phonon coupling, which is needed to describe superconductivity, is *a priori* not included in standard DFT. However, an extended density functional theory including electron-phonon coupling can be formulated, also allowing the description of superconductors [63]. The list of limitations is by no means complete and great care needs to be taken with respect to the applicability of any computational method on a case to case basis.

The scope of this work is to describe laser induced defect structures with computational means and to understand laser damage in CaF_2 on a microscopic level. In chapter 2 I will outline the basic theory of the computational methods I used. Chapter 3 discusses the properties of vacancies (F centers) and interstitials (H centers) in CaF_2 and their formation via the self-trapped exciton. In addition, the stabilization of these point defects by impurities, but also the self-stabilization of the F centers is discussed. The diffusion properties of the point defects are of importance for the formation of larger defect structures, namely metallic Ca colloids, whose optical properties are described with Mie-theory. Chapter 4 gives a summary of the most important experimental results with respect to laser damage. Further, I explain the laser damage model I have developed in chapter 5, providing an understanding of the damaging mechanism, the reversibility of laser damage and suggestions for the prevention of laser damage. Finally, I summarize the outcome of my work and discuss the outlook and future perspectives.

Chapter 2

Calculation of material properties

This chapter revises the basic theory for the calculation of material properties based on density functional theory. The computation of special properties like diffusion and vibrational modes are discussed. For the underlying theory, I will follow the illustration by Martin [64]. An overview and a very good introduction to density functional theory is also given in [65].

2.1 Describing Solid State Matter

Approaching solid state matter from a theoretical point of view is challenging because many aspects need to be taken into account. Solid state matter is a many-particle system whose macroscopic properties are described by the laws of thermodynamics. These properties are largely determined by the electronic structure, which requires a quantum mechanical description.

The thermodynamic approach

In a standard thermodynamic approach, any macroscopic physical system in thermodynamic equilibrium can be described as an ensemble of particles, which is characterized by its equation of state, given by

$$dU = TdS - pdV + \mu dN + \ldots \tag{2.1}$$

for a closed system. For a stable structure in equilibrium, all other macroscopic properties can be derived from it. If the internal energy $U(S, V, N)$ was known, the system would be determined completely.

The properties of solid state matter are mainly determined by its electronic properties, for which quantum mechanical effects such as exchange and correlation need to be considered. Therefore, the description of solid state matter relies on determining the total energy from the quantum mechanical Hamiltonian.

The quantum mechanical approach

Every quantum mechanical system is fully characterized by its Hamiltonian. Any material can be considered as a system of nuclei and electrons described by the general Hamiltonian

$$\begin{aligned} H = &-\frac{\hbar^2}{2m_e}\sum_i \nabla_i^2 - \sum_{i,I}\frac{Z_I e^2}{|\mathbf{r}_i - \mathbf{R}_I|} + \frac{1}{2}\sum_{i\neq j}\frac{e^2}{|\mathbf{r}_i - \mathbf{r}_j|} \\ &- \sum_I \frac{\hbar^2}{2M_I}\nabla_I^2 + \frac{1}{2}\sum_{I\neq J}\frac{Z_I Z_J e^2}{|\mathbf{R}_I - \mathbf{R}_J|}, \end{aligned} \quad (2.2)$$

where lower and upper case indices refer to electrons and nuclei respectively, $\mathbf{r}_i$ and $\mathbf{R}_I$ are the electron and nuclei coordinates respectively, e and m_e are the electron charge and the electron mass and Z_I and M_I are the charge and the mass of the nuclei respectively. Besides the kinetic energy of the electrons and the nuclei, it contains the Coulomb interaction between the electrons, the nuclei, and between electrons and nuclei.

The Born-Oppenheimer approximation

The Born-Oppenheimer or adiabatic approximation [66] assumes that the dynamics of the electrons does not effect the dynamics of the nuclei due to inertia, which in turn implies that the electron system reacts instantly to any movement of the nuclei.

The term proportional to $1/M_I$ in the general Hamiltonian (2.2) can be regarded as small because the nucleon mass is larger than the electron mass by three orders of magnitude and a perturbation series can be defined in terms of this parameter. The actual approximation is to set the nucleon mass to infinity, leaving the nuclei as a static system, which can be seen as an external potential for the electrons. The full system is thus reduced to an electron-only many-body system with an external potential, resulting in the so-called Born-Oppenheimer Hamiltonian

$$H = T + V_{\text{ext}} + V_{\text{int}} + E_{II}, \quad (2.3)$$

where the kinetic energy operator for the electrons T is given by

$$T = -\frac{\hbar}{2m_e}\sum_i \nabla_i^2, \quad (2.4)$$

the interaction between the nuclei and the electrons is described by

$$V_{\text{ext}} = \sum_{i,I}\frac{Z_I e^2}{|\mathbf{r}_i - \mathbf{R}_I|}, \quad (2.5)$$

2.1 Describing Solid State Matter

V_{int} is the electron-electron interaction

$$V_{\text{int}} = \frac{1}{2} \sum_{i \neq j} \frac{e^2}{|\mathbf{r}_i - \mathbf{r}_j|}, \qquad (2.6)$$

and E_{II} describes the classical interaction of the nuclei with one another

$$E_{II} = \frac{1}{2} \sum_{I \neq J} \frac{Z_I Z_J e^2}{|\mathbf{R}_I - \mathbf{R}_J|}. \qquad (2.7)$$

The Born-Oppenheimer approximation decouples the nucleonic dynamics from the electronic dynamics. This implies that a static electron-phonon coupling as in the Jahn-Teller effect [67] is still accessible, but a dynamic electron-phonon coupling as in superconductors [68] is not[1].

The Schrödinger equation

The dynamics of the electrons is described by the time-dependent Schrödinger equation,

$$i\hbar \frac{\mathrm{d}\Psi(\{\mathbf{r}_i\}, t)}{\mathrm{d}t} = \hat{H}\Psi(\{\mathbf{r}_i\}, t), \qquad (2.8)$$

where $\Psi(\{\mathbf{r}_i\}, t) \equiv \Psi(\mathbf{r}_1, \mathbf{r}_2, \ldots, \mathbf{r}_N, t)$ is the many-electron wave function and the spin is assumed to be included in the coordinate $\mathbf{r}_i$. The eigenstates of the system can be written as $\Psi(\{\mathbf{r}_i\}, t) = \Psi(\{\mathbf{r}_i\}) e^{-i(E/\hbar)t}$.

The density of particles $n(\mathbf{r})$ is a central quantity in electronic structure theory and is given by the expectation value of the density operator $\hat{n}(\mathbf{r}) = \sum_{i=1,N} \delta(\mathbf{r} - \mathbf{r}_i)$,

$$n(\mathbf{r}) = \frac{\langle \Psi | \hat{n}(\mathbf{r}) | \Psi \rangle}{\langle \Psi | \Psi \rangle} = N \frac{\int \mathrm{d}^3 r_2 \cdots \mathrm{d}^3 r_N \sum_\sigma |\Psi(\mathbf{r}, \mathbf{r}_2, \ldots, \mathbf{r}_N)|^2}{\int \mathrm{d}^3 r_1 \mathrm{d}^3 r_2 \cdots \mathrm{d}^3 r_N |\Psi(\mathbf{r}_1, \mathbf{r}_2, \ldots, \mathbf{r}_N)|^2}, \qquad (2.9)$$

where the sum is over all spin states σ.

The total energy is the expectation value of the Hamiltonian,

$$E = \frac{\langle \Psi | \hat{H} | \Psi \rangle}{\langle \Psi | \Psi \rangle} \equiv \langle \hat{H} \rangle = \langle \hat{T} \rangle + \langle \hat{V}_{\text{int}} \rangle + \int \mathrm{d}^3 r V_{\text{ext}}(\mathbf{r}) n(\mathbf{r}) + E_{II}, \qquad (2.10)$$

where the external potential can be expressed by an explicit integral over the density. The eigenstates of the Hamiltonian can be obtained from the variational principle and must fulfill

---

[1] The Jahn-Teller effect describes a static lattice distortion (i.e. a phonon) for a system with a degenerate ground state, which removes the degeneracy and leads to a lower overall energy of the system. In superconductors the electrons explicitly couple to the dynamics of the nuclei, which is known as Fröhlich coupling.

the time-independent Schrödinger equation

$$\hat{H}|\Psi\rangle = E|\Psi\rangle. \tag{2.11}$$

The electronic structure

The electronic structure is fully described by the Born-Oppenheimer Hamiltonian. The solution of the Schrödinger equation yields the energy eigenvalues and eigenstates, from which the density of states can be derived. Since electrons are fermions, the energy eigenstates are occupied according to Fermi-Dirac statistics. In large-bandgap materials it is legitimate to neglect any temperature or entropic effects of the electrons because the thermal energy $k_B T$ at "normal" temperatures is much smaller than electronic energies (at room-temperature $k_B T = 25$ meV, the bandgap of CaF_2 is 11.8 eV). However, in metals and semiconductors the electronic contribution to the thermodynamical properties is of great importance. Since CaF_2 has a large bandgap I do not consider any electronic contributions to thermodynamical properties and it is sufficient to calculate the electronic structure at $T = 0$.

The dynamics of the nuclei

The dynamics of the nuclei can be reintroduced as an external perturbation of the system. Under the assumption that the nuclei move in a certain potential, e.g. the harmonic approximation, the system is perturbed by a small displacement of the nuclei, yielding the phonon modes as vibrational eigenstates of the system and the phononic density of states ("frozen phonon" approach). Alternatively, these quantities can be obtained from a linear response approach. The calculation of phonon frequencies is discussed in chapter 2.3.3.

Since typical phonon energies are in the THz (meV) range, which is the range of room-temperature, the phononic density of states can be occupied according to Bose-Einstein statistics, and the thermodynamic properties such as the entropy can be deduced from the quantum mechanical calculation. Since the thermal occupation of the electron states does not play any role in large-bandgap materials, the thermodynamic properties only depend on the dynamics of the nuclei.

2.2 Density Functional Theory

Density functional theory (DFT) is one method to solve the many-body problem of nuclei-electron systems. It is based on the principle that any property of a system of interacting particles can be written as a functional of the ground state electron density. The brilliance of DFT is that all quantities are derived from one scalar function of spacial coordinates, the particle density, instead of the wave functions.

2.2 Density Functional Theory

2.2.1 The fundamental variables

For the theoretical description, the total energy at $T = 0$ and a fixed particle number as a function of the volume V is the most practical quantity because it is easier to treat the volume as a parameter than the pressure.

The fundamental quantities are thus given by the energy E and its derivatives the pressure P and the bulk modulus B

$$\begin{aligned} E &= E(\mathrm{V}) \equiv E_{\text{total}}(\mathrm{V}), \\ P &= -\frac{dE}{dV}, \\ B &= -\mathrm{V}\frac{dP}{dV} = \mathrm{V}\frac{d^2 E}{dV^2}, \end{aligned} \quad (2.12)$$

and higher derivatives of the energy. The equilibrium volume V_0 and the bulk modulus B are an important first test, because these quantities can be measured experimentally with great accuracy.

For the relaxation of a structure to its ground state configuration with respect to the atom positions, the forces acting on the single atoms need to be minimized. In accordance with (2.12) the forces and the force constants result as

$$\begin{aligned} E &= E(\{\mathbf{R}_I\}) \equiv E_{\text{total}}(\{\mathbf{R}_I\}), \\ \mathbf{F}_I &= -\frac{dE}{d\mathbf{R}_I}, \\ C_{IJ} &= -\frac{d\mathbf{F}_I}{d\mathbf{R}_J} = \frac{d^2 E}{d\mathbf{R}_I d\mathbf{R}_J}, \end{aligned} \quad (2.13)$$

where the Born-Oppenheimer approximation is assumed. Minimizing the forces $\mathbf{F}_I$ yields the structure-relaxed ground state of the system. The force constants C_{IJ} become important for the calculation of the vibrational properties.

2.2.2 The Hohenberg-Kohn theorems

Modern DFT is based on two theorems by Hohenberg and Kohn [51] which established the particle density as the fundamental quantity for many-body systems. Hohenberg and Kohn formulate DFT as an exact theory of many-body systems with a Hamiltonian, which can be written as

$$\hat{H} = -\frac{\hbar^2}{2m_e}\sum_i \nabla_i^2 + \sum_i V_{\text{ext}}(\mathbf{r}_i) + \frac{1}{2}\sum_{i \neq j}\frac{e^2}{|\mathbf{r}_i - \mathbf{r}_j|}. \quad (2.14)$$

- **Theorem I:** The ground state particle density $n_0(\mathbf{r})$ of a many-body system uniquely determines the external potential V_{ext} up to a constant.

Corollary I: Any property of the system can be written as a function of $n_0(\mathbf{r})$ because the Hamiltonian and therefore also the wave functions are fully characterized.

- **Theorem II:** The energy functional $E[n]$ can be defined for any potential V_{ext}. The global minimum of this functional is the ground state energy, and the density that minimizes the functional is the exact ground state density $n_0(\mathbf{r})$.

  Corollary II: The functional $E[n]$ by itself determines the ground state energy and ground state density.

According to these two theorems the total energy of a system can be written as a functional of the ground state density

$$\begin{aligned} E_{\text{HK}}[n] &= T[n] + E_{\text{int}}[n] + \int d^3 r V_{\text{ext}}(\mathbf{r}) n(\mathbf{r}) + E_{II} \\ &\equiv F_{\text{HK}}[n] + \int d^3 r V_{\text{ext}}(\mathbf{r}) n(\mathbf{r}) + E_{II}, \end{aligned} \quad (2.15)$$

where F_{HK} includes all internal energies of the interacting electron system, V_{ext} is the external potential imposed by the nuclei, and E_{II} is the interaction energy of the nuclei.

The reformulation of the many-body problem in terms of functionals of the particle density is the achievement of the Hohenberg-Kohn theorems. However, they merely state that any property can be expressed as a functional of the density.

2.2.3 The Kohn-Sham approach

Kohn and Sham [52] introduced the idea to replace the complex interacting many-body system of Hohenberg-Kohn with an auxiliary system, which allows for an easier solution. The auxiliary system is to be chosen in a way, that its ground state density represents the ground state density of the full interacting system. The Hamiltonian of the auxiliary system is constructed with the usual kinetic operator and some effective local potential. The reformulation is in principle exact, however the exact effective potential, i.e. the exchange-correlation functional, is not known explicitly and can only be approximated.

The Kohn-Sham Ansatz

In practice the Hamiltonian of the auxiliary independent-particle system can be written as

$$\hat{H}_{\text{aux}}^{\sigma} = -\frac{1}{2}\nabla^2 + V^{\sigma}(\mathbf{r}), \quad (2.16)$$

where V^{σ} is an effective potential which will be specified later. For a system of $N = N^{\uparrow} + N^{\downarrow}$ independent electrons the ground state has one electron in each of the N^{σ} orbitals $\psi_i^{\sigma}(\mathbf{r})$ with

2.2 Density Functional Theory

the lowest eigenvalues ϵ_i^σ, which means the ground state density can be written as

$$n(\mathbf{r}) = \sum_\sigma n(\mathbf{r},\sigma) = \sum_\sigma \sum_{i=1}^{N^\sigma} |\psi_i^\sigma(\mathbf{r})|^2. \qquad (2.17)$$

In principle the quasi single-particle wave functions ψ_i^σ do not have any physical meaning, but in practice their interpretation as physical states works very well.

The independent-particle kinetic energy T_S is given by

$$T_S = -\frac{1}{2} \sum_\sigma \sum_{i=1}^{N^\sigma} \langle \psi_i^\sigma | \nabla^2 | \psi_i^\sigma \rangle = \frac{1}{2} \sum_\sigma \sum_{i=1}^{N^\sigma} \int d^3r |\nabla \psi_i^\sigma(\mathbf{r})|^2, \qquad (2.18)$$

and the classical Coulomb interaction energy is defined as

$$E_{\text{Hartree}}[n] = \frac{1}{2} \int d^3r d^3r' \frac{n(\mathbf{r})n(\mathbf{r}')}{|\mathbf{r}-\mathbf{r}'|}. \qquad (2.19)$$

The Kohn-Sham approach then proceeds by rewriting the Hohenberg-Kohn expression for the total energy functional (2.15) as

$$E_{\text{KS}} = T_S[n] + \int d\mathbf{r} V_{\text{ext}}(\mathbf{r}) n(\mathbf{r}) + E_{\text{Hartree}}[n] + E_{II} + E_{\text{xc}}[n]. \qquad (2.20)$$

Comparing the Kohn-Sham and the Hohenberg-Kohn total energy functionals (2.20 and 2.15), the exchange-correlation functional can be written as

$$E_{\text{xc}}[n] = T[n] - T_S[n] + E_{\text{int}}[n] - E_{\text{Hartree}}[n], \qquad (2.21)$$

which is just the difference of the kinetic and potential energy of the interacting system and the auxiliary non-interacting system with the electron-electron interaction replaced by the Hartree energy. If the functional E_{xc} was known, then the exact ground state energy and density of the system could be calculated by solving the Kohn-Sham equations, because all other terms in the Kohn-Sham total energy functional (2.20) can be calculated exactly up to the accuracy of numerical methods.

The Kohn-Sham equations

The Kohn-Sham auxiliary system can be solved using the variational principle. Applying the variational principle to (2.20), results to

$$\frac{\delta E_{\text{KS}}}{\delta \psi_i^{\sigma*}(\mathbf{r})} = \frac{\delta T_S}{\delta \psi_i^{\sigma*}(\mathbf{r})} + \left[\frac{\delta E_{\text{ext}}}{\delta n(\mathbf{r},\sigma)} + \frac{\delta E_{\text{Hartree}}}{\delta n(\mathbf{r},\sigma)} + \frac{\delta E_{\text{xc}}}{\delta n(\mathbf{r},\sigma)} \right] \frac{\delta n(\mathbf{r},\sigma)}{\delta \psi_i^{\sigma*}(\mathbf{r})} = 0, \qquad (2.22)$$

where the orthonormalization constraints are

$$\langle \psi_i^\sigma | \psi_j^{\sigma'} \rangle = \delta_{i,j} \delta_{\sigma,\sigma'}. \tag{2.23}$$

Using (2.17) and (2.18) leads to

$$\frac{\delta T_S}{\delta \psi_i^{\sigma*}(\mathbf{r})} = -\frac{1}{2} \nabla^2 \psi_i^\sigma(\mathbf{r}) \quad \text{and} \quad \frac{\delta n^\sigma(\mathbf{r})}{\delta \psi_i^{\sigma*}(\mathbf{r})} = \psi_i^\sigma(\mathbf{r}). \tag{2.24}$$

With the Langrange multiplier method for constraints one obtains the Schrödinger-like Kohn-Sham equations

$$(H_{\text{KS}}^\sigma - \epsilon_i^\sigma) \psi_i^\sigma(\mathbf{r}) = 0, \tag{2.25}$$

which form a system of coupled single-particle equations. Here ϵ_i^σ are the eigenvalues, and H_{KS} is the effective Hamiltonian

$$H_{\text{KS}}^\sigma(\mathbf{r}) = -\frac{1}{2} \nabla^2 + V_{\text{KS}}^\sigma(\mathbf{r}), \tag{2.26}$$

with

$$\begin{aligned} V_{\text{KS}}^\sigma(\mathbf{r}) &= V_{\text{ext}}(\mathbf{r}) + \frac{\delta E_{\text{Hartree}}}{\delta n(\mathbf{r}, \sigma)} + \frac{\delta E_{\text{xc}}}{\delta n(\mathbf{r}, \sigma)} \\ &= V_{\text{ext}}(\mathbf{r}) + V_{\text{Hartree}}(\mathbf{r}) + V_{\text{xc}}^\sigma(\mathbf{r}). \end{aligned} \tag{2.27}$$

Up to this point these equations and therefore the solution of the many-particle system is independent of any approximations, i.e. the system would have an exact solution if the functional $E_{\text{xc}}[n]$ was known. Since the Hohenberg-Kohn theorems state that the ground state density determines the potential at the minimum, a unique Kohn-Sham potential $V_{\text{KS}}^\sigma(\mathbf{r}) \equiv V_{\text{eff}}^\sigma(\mathbf{r})|_{\text{min}}$ can be associated with any given interacting electron system.

The great advantage of the Kohn-Sham approach is that the kinetic energy, the external potential and the long-range Hartree-terms are separated from the exchange and correlation terms. The first three can be calculated exactly depending only on the accuracy of the numerical method while the exchange-correlation functional can be approximated by a local functional of the density. The exchange-correlation energy E_{xc} can be expressed as

$$E_{\text{xc}}[n] = \int d\mathbf{r}\, n(\mathbf{r})\, \epsilon_{\text{xc}}([n], \mathbf{r}), \tag{2.28}$$

where $\epsilon_{\text{xc}}([n], \mathbf{r})$ is an energy per electron at point $\mathbf{r}$ which only depends on the density $n(\mathbf{r}, \sigma)$ in a certain neighborhood of point $\mathbf{r}$.

2.2.4 Exchange-correlation functionals

The exchange-correlation functional in the Kohn-Sham approach is *a priori* an unknown functional. However, finding a good, yet simple approximation for the complex exchange-correlation functional is crucial for the correct description of the electronic properties.

The local density approximation (LDA)

The local density approximation (LDA), or more generally the local spin density approximation (LSDA), is based on the consideration that solids with a slowly varying electron density are oftentimes close to the limit of the homogeneous electron gas [52]. Therefore, it is a valid approximation to assume that the exchange-correlation energy density is the same as in a homogeneous electron gas with the same density, which leads to the exchange-correlation energy functional

$$\begin{aligned} E_{\text{xc}}^{\text{LSDA}}[n^\uparrow, n^\downarrow] &= \int d^3r \, n(\mathbf{r}) \, \epsilon_{\text{xc}}^{\text{hom}}(n^\uparrow(\mathbf{r}), n^\downarrow(\mathbf{r})) \\ &= \int d^3r \, n(\mathbf{r}) [\epsilon_{\text{x}}^{\text{hom}}(n^\uparrow(\mathbf{r}), n^\downarrow(\mathbf{r})) + \epsilon_{\text{c}}^{\text{hom}}(n^\uparrow(\mathbf{r}), n^\downarrow(\mathbf{r}))]. \end{aligned} \qquad (2.29)$$

Since LDA assumes a more or less homogeneous electron distribution it cannot describe rapid variations of the electron density correctly. While LDA should work well for metals, it is expected to fail for ionic solids and materials with strongly localized electrons.

Generalized-gradient approximations (GGAs)

It is straight forward to assume that the description of systems with a rapidly varying electron density can be improved by including the gradient of the density in the exchange-correlation functional [52]. However, a simple gradient expansion does not lead to a consistent improvement for various reasons [69]. Therefore, several general gradient approximations have been proposed [58, 59, 60]. Their starting point is the exchange-correlation functional written as

$$\begin{aligned} E_{\text{xc}}^{\text{GGA}}[n^\uparrow, n^\downarrow] &= \int d^3r \, n(\mathbf{r}) \, \epsilon_{\text{xc}}(n^\uparrow, n^\downarrow, |\nabla n^\uparrow|, |\nabla n^\downarrow|, \ldots) \\ &\equiv \int d^3r \, n(\mathbf{r}) \, \epsilon_x^{\text{hom}}(n) F_{\text{xc}}(n^\uparrow, n^\downarrow, |\nabla n^\uparrow|, |\nabla n^\downarrow|, \ldots). \end{aligned} \qquad (2.30)$$

The explicit inclusion of the gradient of the density has led to a significant improvement with respect to LDA calculations. It was the implementation of GGAs that provided the necessary accuracy, which established DFT as an accepted method in chemistry and materials science.

However, for magnetic materials or compounds with transition metals or rare-earth elements, GGA oftentimes cannot provide correct results. The main problems occur due to strongly localized electrons in $3d$ or $4f$ shells and strong electron correlation. The description of such materials requires more advanced functionals or even a full solution of the strongly interacting many-particle system, which is not possible for more than three particles.

Orbital-dependent and hybrid functionals

Since there is no systematic approach to improve the exchange-correlation functionals within the Kohn-Sham approach, new or altered functionals are needed for the description of materials and compounds, where $3d$ and $4f$ electrons play a significant role. These are mainly transition metals and rare-earth elements and their compounds. The two main approaches to describing these materials correctly are orbital-dependent functionals or hybrid functionals. An example for an orbital-dependent functional is LDA+U or GGA+U, which is based on LDA or GGA and has a Hubbard-like U-interaction added to account for exchange and correlation due to strongly localized electrons by considering on-site repulsion or attraction. One example for hybrid functionals is the combination of orbital-dependent Hartree-Fock and an explicit density functional as implemented in the B3LYP functional, which combines the GGA exchange functional B88 [58] with the orbital-dependent LYP [62] correlation.

2.2.5 Pseudopotentials

In principle the Hamiltonian in equation (2.14) describes all electrons in a given material. However, only the electrons in the outer shells, the so-called valence electrons effectively contribute to the electronic properties of the material, as the core electrons are closely bound to the nucleus. To reduce the computational effort, it is desirable to effectively only calculate the valence electrons. In addition, one is strongly interested in reducing the number of basis functions necessary for the expansion of the wave function, since the computational effort for solving the Schrödinger equation scales with the third power of the number of basis functions.

With the introduction of so-called pseudopotentials the full atom potentials are replaced by effective potentials that include the core electrons in a core region, leaving only the valence electrons for the explicit solving of the Schrödinger equation. These pseudopotentials are generated from an all-electron calculation of the atom, matching the pseudo wave function and its derivatives to the all-electron wave function and its derivatives outside a certain core radius R_c. Since the all-electron wave function is calculated with *ab initio* methods, no empirical parameters are needed for the construction of the pseudopotentials. For norm-conserving pseudopotentials the integrated charge inside R_c also agrees for both wave functions. The effect of using pseudopotentials is that the resulting wave functions are smoother, which results in a lower number of basis functions necessary to describe the system with the desired accuracy.

It is important to keep in mind that the choice of pseudopotentials is crucial for the outcome of any calculation. Even though the quality and versatility of pseudopotentials has been improved in recent years, not every pseudopotential is suitable for the solution of every problem.

Ultrasoft pseudopotentials

Making the pseudofunctions as smooth as possible without loosing accuracy is one goal of pseudopotentials, because increasing the smoothness of the pseudopotentials decreases the range in Fourier space necessary to describe the valence properties to a certain accuracy. In Vanderbilt's Ultrasoft Pseudopotentials (USPP) [70] the smoothness of the pseudofunctions in comparison to norm-conserving pseudopotentials is increased by giving up the restriction of norm conservation. This led to much smaller cutoff energies for the expansion of the wave function in k-space resulting in faster computing times while maintaining the desired accuracy.

Projector-augmented waves (PAW)

The projector-augmented wave (PAW) method [71, 72] combines the pseudopotential approach with an all-electron approach. The wave function is parametrized by a pseudofunction, but in the core region the pseudized wave function is substituted with the all-electron wave function. This combines the great advantages of USPP, namely the reduction of the computational effort with the accuracy of an all-electron method.

2.2.6 Wave function expansion

In principle, the wave functions in the particle-independent Kohn-Sham equations (2.25) can be expanded in any complete system of orthonormal basis functions. However, it is convenient to choose the wave function expansion according the boundary conditions and the symmetries of the system. For solid state materials with a periodic structure the plane wave expansion has the great advantage, that it automatically satisfies the periodic boundary conditions. When calculating molecules without periodic boundary conditions or surfaces that do not fill the space of the simulation cell, it can be more convenient to use local basis functions, such as Gaussians or spherical harmonics.

2.3 Calculating diffusion properties

In general, the diffusion coefficient is written in the form

$$D = D_0 \, e^{-\frac{\Delta E}{k_B T}}, \qquad (2.31)$$

where ΔE is the barrier height, D_0 is a prefactor, k_B the Boltzmann constant, and T the absolute temperature. For practical reasons I will use the formulation by Wert and Zener [73]

$$D = n\beta d^2 \Gamma \quad \text{with} \quad \Gamma = \Gamma_0 \, e^{-\frac{\Delta E}{k_B T}}, \qquad (2.32)$$

where n is the number of stable nearest-neighbor sites for the diffusing atom, β is the probability that a jump leads forward in the direction of diffusion, and d is the length of the jump projected onto the direction of diffusion. Γ represents the jump rate between adjacent sites.

The calculation of diffusion properties is based on Eyring's transition state theory [74]. A static approach to determining the diffusion path is the nudged elastic band method, which assumes an initial path and optimizes it leading to the nearest "downhill" minimum. The method is therefore dependent on the choice of the initial path.

Another approach is to sample the diffusion path with classical MD or CPMD and statistically determine the diffusion properties from a large number of events. This method is limited to systems with small barriers because a large number of observed diffusion events is necessary for statistical reasons and the probability of observing events decreases rapidly with an increasing barrier.

Transition path sampling [75] is a more advanced approach based on a Monte-Carlo approach with random path alterations evaluating each path's probability. It is designed to describe rare events on rugged, complex energy surfaces, which are not accessible with the dynamic approach.

Since I expect barrier heights of up to 2 eV and the energy surface in CaF_2 to be rather smooth, I will use the nudged elastic band method to calculate the diffusion properties of defects in CaF_2. On the one hand the expected barriers are too high to obtain a sufficient number of events with dynamical sampling, and on the other hand the simplicity of the highly symmetric CaF_2 does not require the computational complexity of transition path sampling.

2.3.1 Transition State Theory

Eyring's basic idea of transition state theory (TST) is that a transition process, i.e. a reaction or diffusion, starts at an initial state, passes a transition state, and ends at a final state [74]. In addition to the Born-Oppenheimer approximation, TST assumes that the transition rate is low enough, that the initial state is characterized by a Boltzmann distribution. The diffusion process can then be described on a $D-1$ dimensional energy hypersurface, which provides a divide between the initial and final state, where D is the number of degrees of freedom of the system. Any trajectory from the initial to the final state crosses the divide only once. The transition state is the lowest point on the divide.

The great challenge is the calculation of the diffusion path between the initial and the final state. The path with the greatest statistical weight is usually the minimum energy path.

2.3 Calculating diffusion properties

At any point along this path the force acting on the atoms is pointing only along this path. The maxima on the minimum energy path are saddle points on the energy hypersurface, i.e. transition states of the diffusion or reaction. If there are more than one transition state, the minimum in between the transition states represents a stable intermediate configuration. The state-of-the-art method for finding transition states and minimum energy paths is the climbing image nudged elastic band method proposed by Henkelmann et al. [76], which I will discuss below.

According to Eyring [74] the reaction or jump rate Γ is given by

$$\Gamma = \frac{k_B T}{h} \frac{Z_{TS}}{Z_i} e^{-\frac{\Delta E_0}{k_B T}}, \qquad (2.33)$$

where h is Planck's constant, Z_{TS} and Z_i are the partition functions of the transition state and the initial state respectively, and ΔE_0 is the energy difference between the initial and transition state at $T = 0$.

According to Wimmer et al. [77] the partition functions can be written as

$$\Gamma = \frac{k_B T}{h} \frac{\prod_{i=1}^{3N-3} \left[2 \sinh\left(\frac{h\nu_i^0}{2k_B T}\right) \right]}{\prod_{i=1}^{3N-4} \left[2 \sinh\left(\frac{h\nu_i^{TS}}{2k_B T}\right) \right]} e^{-\frac{\Delta E_0}{k_B T}} \qquad (2.34)$$

within the framework of the harmonic approximation. ν_i^0 and ν_i^{TS} are the phonon frequencies of the initial and the transition state respectively. The three acoustic phonon modes cancel because they are the same for the initial and the transition state, and the partition function for the transition state is evaluated on the $D-1$ dimensional hypersurface, therefore yielding one phonon mode less than the initial state. In practice, one imaginary phonon frequency is obtained for the transition state, whose eigenvector points along the minimum energy path at the transition state, and thus indicates the direction of the diffusion trajectory at this point. For the high temperature limit, i.e. $h\nu/2k_B T \ll 1$ the jump rate reduces to [78]

$$\Gamma = \frac{\prod_{i=1}^{N-3} \nu_i^0}{\prod_{i=1}^{N-4} \nu_i^{TS}} e^{-\frac{\Delta F}{k_B T}}, \qquad (2.35)$$

where the barrier $\Delta F = \Delta E_0 - T\Delta S$ is given in terms of the free energy.

The crucial part for the calculation of diffusion properties is to find the transition state. With the transition state the energy barrier is determined, and the phonon modes for the initial and transition state can be calculated.

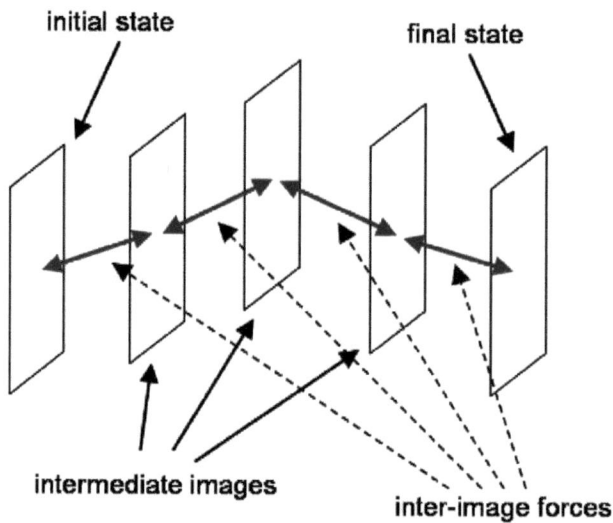

Figure 2.1: Schematic illustration of the nudged elastic band method. The diffusion path is sampled by intermediate images, which are connected with virtual spring forces between the images to keep them distributed along the path.

2.3.2 The climbing image nudged elastic band method

As mentioned above, the climbing image nudged elastic band method, introduced by Henkelmann et al. [76], is the method of choice for determining transition states based on transition state theory. I will follow their description of the method, which is schematically depicted in figure 2.1.

I start off by sampling an initial guess of the diffusion path by a set of $N+1$ discrete images $\{I_0, I_1, \ldots, I_N\}$, where the initial and final state are denoted by I_0 and I_N respectively and are fixed. Adjacent images are attached to each other by virtual spring forces

$$\mathbf{F}_i^s|_\| = k(|I_{i+1} - I_i| - |I_i - I_{i-1}|)\hat{\tau}_i, \qquad (2.36)$$

where k is the arbitrarily chosen spring constant and $\hat{\tau}_i$ is the normalized local tangent to the path at image i. The springs constitute the so-called elastic band from the initial to the final state and keep the images equidistant. It is convenient to define a generalized diffusion coordinate along the elastic band, i.e. along the resulting minimum energy path. Of the true forces acting upon the images only the part perpendicular to the path is considered,

$$-\nabla E_i(\mathbf{R}_j)|_\perp = -\nabla E_i(\mathbf{R}_j) + \nabla E_i(\mathbf{R}_j) \cdot \hat{\tau}_i. \qquad (2.37)$$

2.3 Calculating diffusion properties

Here $E_i(\mathbf{R}_j)$ is the energy of the ith image as a function of all the atomic coordinates $\mathbf{R}_j$. This results in a total force

$$\mathbf{F}_i = \mathbf{F}_i^s|_\parallel - \nabla E_i(\mathbf{R}_j)|_\perp, \tag{2.38}$$

which needs to be evaluated for each image.

In practice, this means that the atom positions in each image are relaxed perpendicular to the diffusion path, while any forces parallel to the path are projected onto the virtual spring forces. This relaxation allows the elastic band to move to the minimum energy path. The projection of the forces needs to be readjusted after every iteration as the path changes with every step.

For the case that the true transition state lies in between two images the climbing image formalism is introduced. The image with the highest energy is identified and then allowed to climb up the path until it reaches the transition state. This is done by inverting the component of the full force which is parallel to the path

$$\begin{aligned}\mathbf{F}_{i_\mathrm{max}} &= -\nabla E_{i_\mathrm{max}}(\mathbf{R}_j) + 2\,\nabla E_{i_\mathrm{max}}(\mathbf{R}_j)|_\parallel \\ &= -\nabla E_{i_\mathrm{max}}(\mathbf{R}_j) + 2\,\nabla E_{i_\mathrm{max}}(\mathbf{R}_j)\cdot\hat{\tau}_{i_\mathrm{max}}\hat{\tau}_{i_\mathrm{max}}.\end{aligned} \tag{2.39}$$

The force on the climbing image is now independent of the spring forces allowing it to move along the path. Since only the force component parallel to the path is inverted the image will climb up the path to the saddle point, while the forces perpendicular to the path ensure that the image stays on the path. The final configuration of the climbing image is the transition state.

It is important to mention that the result of the climbing image nudged elastic band method depends on the initial choice for the diffusion path. From the starting path the method will only find the local "downhill" minimum of the path. It is therefore highly recommendable to chose different starting paths and compare the results in order to make sure that the true minimum energy path is found. While the initial guess for diffusion paths in solids is oftentimes reliable, one should not be intrigued by guessing a path simply for symmetry reasons. The explicit breaking of the symmetry by the choice of path could possibly lead to a lower diffusion barrier. Especially for surface diffusion or chemical adsorption processes, it is important to compare several starting paths to determine the true minimum energy path.

Testing the transition state

A good test for the transition state is the calculation of the corresponding phonon modes. Since the transition state is a saddle point on the energy hypersurface it must have one and only one imaginary phonon mode according to Eyring [74]. The number of imaginary phonon modes can therefore be used as a test whether a true transition state has been found.

Oftentimes the direct interpolation between initial and final state or the initial (human) guess for the diffusion path exhibits a higher symmetry than the true diffusion path, which is the minimum energy path. This can lead to a transition state which is not the true transition state.

The nudged elastic band method is converged when

$$\mathbf{F}_i|_\perp = 0 \quad \text{and} \quad \mathbf{F}_{\text{TS}}|_\parallel = 0. \tag{2.40}$$

However, $\vec{F}_i|_\perp$ can be zero because the path is along a saddle on the energy hypersurface due to symmetry effects. In this case the transition state will exhibit more than one imaginary phonon mode, one along the resulting diffusion path and one or more perpendicular to this diffusion path. The eigenvectors of the imaginary modes perpendicular to the diffusion path can then be used as an indicator to find the true transition state and the minimum energy path.

For the transition state all forces $\mathbf{F}_i(\{\mathbf{R}_{\text{TS}}\}) = 0$, where $\{\mathbf{R}_{\text{TS}}\}$ is the configuration of the transition state obtained from the nudged elastic band calculation. One of the eigenvectors of the imaginary modes will point along the expected diffusion path, the eigenvectors of the other imaginary modes then indicate the displacement towards the minimum energy path. Therefore,

$$\mathbf{F}_i(\{\mathbf{R}_{\text{TS}} + \epsilon\, \mathbf{d}_{\text{im}}\}) \neq 0, \tag{2.41}$$

where ϵ is an arbitrary small parameter and $\mathbf{d}_{\text{im}}$ is the eigenvector corresponding to the imaginary phonon mode perpendicular to the expected diffusion path. Rerunning the nudged elastic band calculation with an interpolated diffusion path that includes the intermediate configuration $\{\mathbf{R}_{\text{TS}} + \epsilon\, \mathbf{d}_{\text{im}}\}$, will then lead to the correct minimum energy path and the true transition state.

In one of my calculations I found a transition state with two imaginary modes, and this procedure worked very well. In principle, this procedure should also work if the transition state resulting from a nudged elastic band calculation exhibits more than two imaginary modes, but I have not been confronted with this situation and could therefore not test it.

2.3.3 Calculating phonon modes

The vibrational or phonon modes can be calculated with the so-called frozen phonon or small displacement method [79] or with response functions in density functional perturbation theory [80].

2.3 Calculating diffusion properties

Lattice dynamics in the Born-Oppenheimer approximation

Within the framework of the Born-Oppenheimer approximation [66], the equation of motion for the nuclei can be written as

$$M_I \frac{\partial^2 \mathbf{R}_I}{\partial t^2} = \mathbf{F}_I(\mathbf{R}) = -\frac{\partial}{\partial \mathbf{R}_I} E(\mathbf{R}), \qquad (2.42)$$

where M_I is the mass of the nuclei, $\mathbf{R} = \{\mathbf{R}_I\}$ is the set of nuclear coordinates, and $E(\mathbf{R})$ is the total electronic energy. The equilibrium position of the nuclei $\mathbf{R}^0$ is defined by $\mathbf{F}_I(\mathbf{R}^0) = 0$. In the harmonic approximation the displacement of the nuclei $\mathbf{u}_I$ is described by

$$\mathbf{u}_I(t) = \mathbf{R}_I(t) - \mathbf{R}_i^0 = \mathbf{u}_I \, e^{-i\omega t} \qquad (2.43)$$

The energy can then be written as a function of the displacements $\mathbf{u}_I$

$$E(\mathbf{u}) = E^0 + \frac{1}{2} \sum_{I,\alpha} \sum_{J,\beta} \underbrace{\frac{\partial^2 E(\mathbf{R})}{\partial R_{I,\alpha} \partial R_{J,\beta}}}_{C_{I,\alpha;J,\beta}} u_{I,\alpha}(t) u_{J,\beta}(t), \qquad (2.44)$$

where $C_{I,\alpha;J,\beta}$ are the inter-atomic force constants with α and β representing Cartesian indices and the equation of motion (2.42) becomes

$$-\omega^2 M_I u_{I,\alpha} = -\sum_{J,\beta} C_{I,\alpha;J,\beta} u_{J,\beta}. \qquad (2.45)$$

The solution to the set of the $3N$ vibrational modes is given by

$$\det \left| \frac{1}{\sqrt{M_I M_J}} C_{I,\alpha;J,\beta} - \omega^2 \right| = 0. \qquad (2.46)$$

The inter-atomic force constants $C_{I,\alpha;J,\beta}$ are calculated in the frozen phonon approach or from linear response theory.

The frozen phonon approach

In the frozen phonon approach, also called the small displacement method, the inter-atomic force constants are determined as the numerical derivative of the forces

$$C_{I,\alpha;J,\beta} \approx -\frac{\Delta \mathbf{F}_{I,\alpha}}{\Delta \mathbf{R}_{J,\beta}}. \qquad (2.47)$$

Every atom is displaced by $\pm \Delta \mathbf{R}_{J,\beta}$ from its equilibrium position yielding $6N$ configurations, for which the forces on all atoms are calculated. For each displacement $\Delta \mathbf{R}_{J,\beta}$ the forces are calculated as the average of the positive and negative displacements, and since the equilibrium forces are zero, the inter-atomic force constants result to

$$C_{I,\alpha;J,\beta} = -\frac{\mathbf{F}_{I,\alpha}(+\Delta\mathbf{R}_{J,\beta}) - \mathbf{F}_{I,\alpha}(-\Delta\mathbf{R}_{J,\beta})}{2\Delta\mathbf{R}_{J,\beta}} \qquad (2.48)$$

The computational effort of this method can be considerably reduced by exploiting the symmetries of the system. With the resulting inter-atomic force constants the phonon modes and their eigenvectors are obtained from the solution of (2.46).

Phonons from linear response

According to the Hellmann-Feynman theorem [81, 82], the forces on the nuclei in the Born-Oppenheimer approximation can be written as

$$\mathbf{F}_I = -\frac{\partial E(\mathbf{R})}{\partial \mathbf{R}_I} = -\left\langle \Psi(\mathbf{R}) \left| \frac{\partial H(\mathbf{R})}{\partial \mathbf{R}_I} \right| \Psi(\mathbf{R}) \right\rangle, \qquad (2.49)$$

where $\Psi(\mathbf{R})$ are the electronic ground state wave functions of the Hamiltonian (2.3-2.7). Thus the forces result to

$$\mathbf{F}_I = -\int d\mathbf{r}\, n(\mathbf{r}) \frac{\partial V_{\text{ext}}(\mathbf{r})}{\partial \mathbf{R}_I} - \frac{\partial E_{II}}{\partial \mathbf{R}_I}. \qquad (2.50)$$

It is straight forward to derive the inter-atomic force constants

$$\begin{aligned} C_{I,\alpha;J,\beta} &= \frac{\partial^2 E(\mathbf{R})}{\partial \mathbf{R}_I \partial \mathbf{R}_J} = -\frac{\partial \mathbf{F}_I}{\partial \mathbf{R}_J} \\ &= \int d\mathbf{r}\, \frac{\partial n(\mathbf{r})}{\partial \mathbf{R}_J} \frac{\partial V_{\text{ext}}(\mathbf{r})}{\partial \mathbf{R}_I} + \int d\mathbf{r}\, n(\mathbf{r}) \frac{\partial^2 V_{\text{ext}}(\mathbf{r})}{\partial \mathbf{R}_I \partial \mathbf{R}_J} + \frac{\partial^2 E_{II}}{\partial \mathbf{R}_I \partial \mathbf{R}_J}. \end{aligned} \qquad (2.51)$$

This approach to calculating the inter-atomic force constants is called linear response because the force constants only depend on the electron density $n(\mathbf{r})$ and its linear response to a lattice distortion $\partial n(\mathbf{r})/\partial \mathbf{R}_I$. Once the inter-atomic force constants are obtained, the phonon modes are again calculated from (2.46).

2.4 Codes and Resources

I used two different codes for the electronic structure calculations.

My primarily used code is the commercially available *Vienna ab-initio simulation package* (VASP) version 5.2 embedded in the MedeA environment (Materials Design Inc.). VASP is a plane wave code and I used projector-augmented wave (PAW) potentials [71, 72] and GGA-PBE [60] exchange-correlation functionals. The calculations of defect structures were based on a $Ca_{32}F_{64}$ supercell.

2.4 Codes and Resources

Second, I used the open source *Quantum Espresso* (QE) package [83]. QE is based on the plane wave *pwscf* code. Here I used ultrasoft pseudopotentials (USPP) [70] and GGA-PBE [60] exchange-correlation functionals. For the nudged elastic band calculations and the phonons from linear response, I used the implementations in QE.

For the calculation of the phonons via the frozen phonon method I used the code PHON [84].

Chapter 3

Properties of defect structures in CaF$_2$

A very good and complete summary of the properties of CaF$_2$ and its point defect structures is given by Hayes [34].

3.1 Pure CaF$_2$

Calcium fluoride, also known as fluorite, has the fluorite structure, i.e. a cubic structure. The fluorite structure consists of cubes with anions at their corners, where every second cube hosts a divalent cation, leaving every other cube empty. The basic structure of CaF$_2$ is shown in figure 3.1. The experimental bandgap of CaF$_2$ is 11.8 eV, my calculations yield a bandgap of 7.3 eV. Due to the approximation of the exchange-correlation functional, the bandgap in DFT calculations is always too small. For large bandgap materials such as CaF$_2$, this difference can be quite large. However, for one of the fundamental quantities, the bulk modulus, I obtain 93.6 GPa matching the experimental value of 83 GPa [85] quite well. The band structure and the total density of states for CaF$_2$ are shown in figure 3.2.

3.2 Point defects in CaF$_2$

3.2.1 The F center

The F center in CaF$_2$ is an electron localized in a fluorine vacancy. The structure is shown in figure 3.3. The electron localization function (ELF) [86] shows that the electron is well localized on the vacancy, which is also confirmed by the spin density, also called the magnetization density, which is the difference of the electron densities for spin up and spin down electrons (figure 3.4). Electronic structure calculations show an occupied state in the band gap without dispersion which corresponds to the localized electron (figure 3.5).

Coloration phenomena of fluorite by the F center have been well investigated by Przibram

3.2 Point defects in CaF$_2$

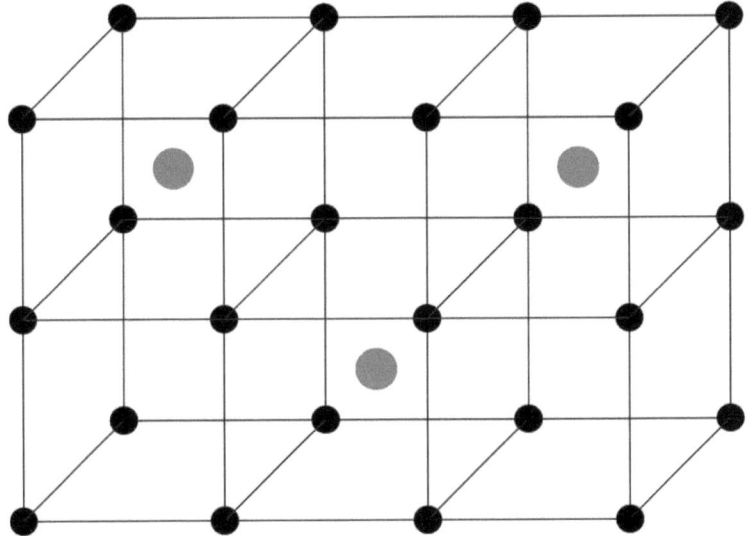

Figure 3.1: The fluorite structure: The anions (F$^-$) occupy the corners of the cubes, every second cube is occupied by a cation (Ca^{2+}) and every other cube is empty.

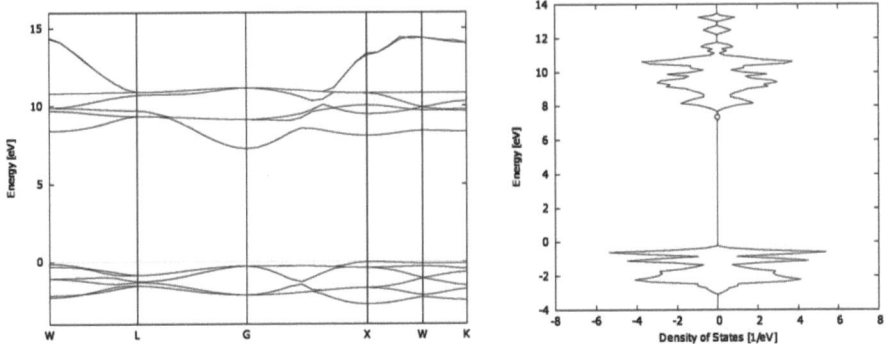

Figure 3.2: Band structure (left) and density of states (right) of pure CaF$_2$.

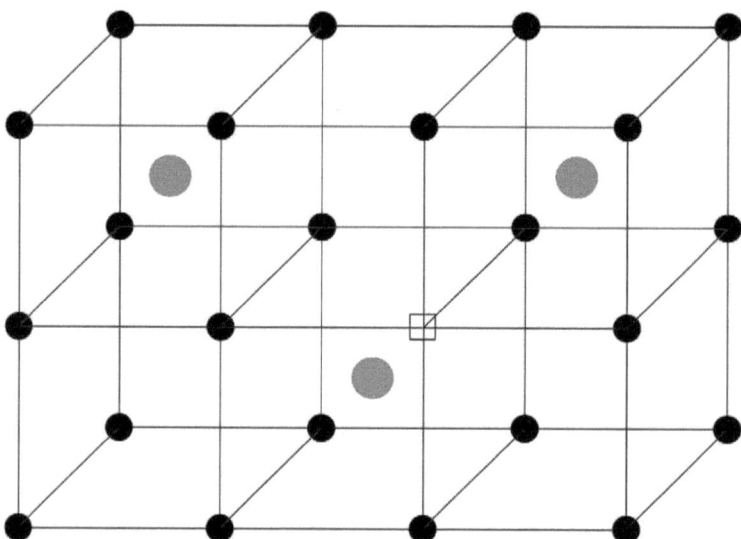

Figure 3.3: The structure of the F center: The electron is located on the fluorine vacancy (square). The calculated fully relaxed structure is shown in figure B.2.

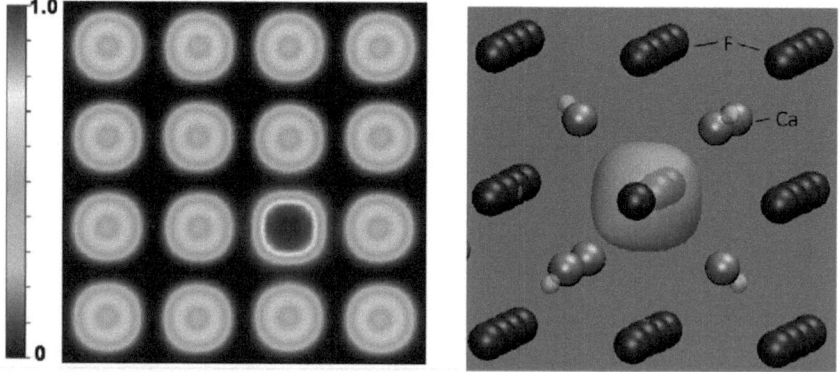

Figure 3.4: The ELF in the [100] plane (left) and the spin density (right) confirm that the electron of the F center is localized in the fluorine vacancy.

3.2 Point defects in CaF_2

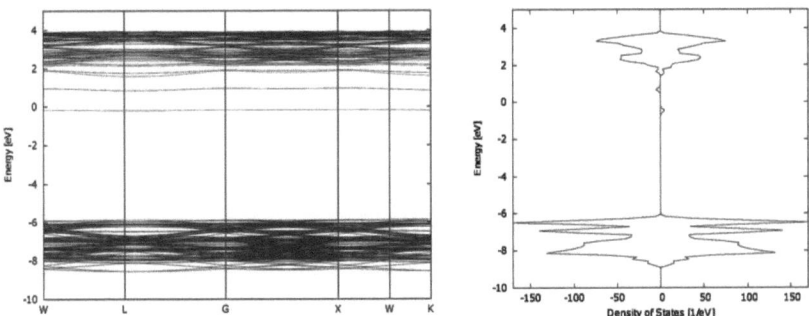

Figure 3.5: Band structure (left) and density of states (right) of the F center: The band structure shows well defined states in the bandgap with no dispersion indicating a well localized defect state.

[23, 24, 25] and Bill [30]. The absorption band of the F-center at 378 nm (3.28 eV) is well-established [87, 88, 89, 90, 91]. Its diffusion properties are calculated in chapter 3.4.

3.2.2 The M center

The M center is a fluorine divacancy, i.e. two adjacent F centers. The structure is shown in figure 3.6. The ELF confirms the strong localization of the electrons in the adjacent vacancies (figure 3.7). Because the M center has only doubly occupied bands (figure 3.8), i.e. paired spin up and spin down electrons, the spin density is not applicable to visualize the localization of the electrons. In correspondence to the F center, the M center exhibits a doubly occupied band in the bandgap (figure 3.8), which matches the band of the F center very well, also not showing dispersion.

The M center shows absorption bands at 365 nm (3.4 eV) and 520 nm (2.4 eV) [92, 90, 93]. Its emission band lies at 600 nm (2.1 eV) [93]. The M center plays an important role in laser damage processes and its formation and influence on the laser stability have been investigated thoroughly [94, 95, 96, 17].

3.2.3 The V_k center

The V_k center in CaF_2 is also referred to as the self-trapped hole. It occurs when the hole of an excited electron-hole pair localizes on two adjacent fluorine ions, which form an F_2^- dimer with an F-F bond along the (100) direction resulting in a slight displacement of the two F ions towards each other [97]. An absorption band of the V_k center is reported at 320 nm (3.87 eV) [90, 98]. The V_k center shows a high mobility in the crystal [99, 100], however it is not stable

Properties of defect structures in CaF_2

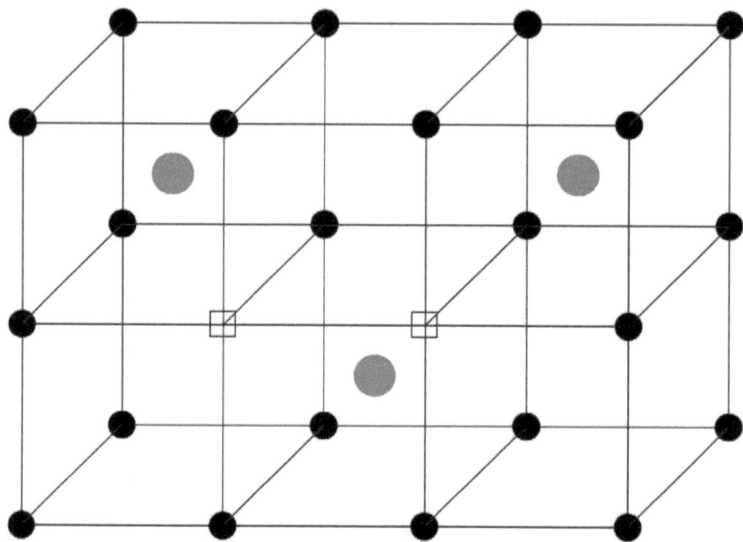

Figure 3.6: The structure of the M center: The two fluorine vacancies (squares) of the M center in CaF_2 are adjacent to each other. A diagonal configuration along the (110) or (111) direction is also possible, but is energetically not favored. The calculated fully relaxed structure is shown in figure B.3.

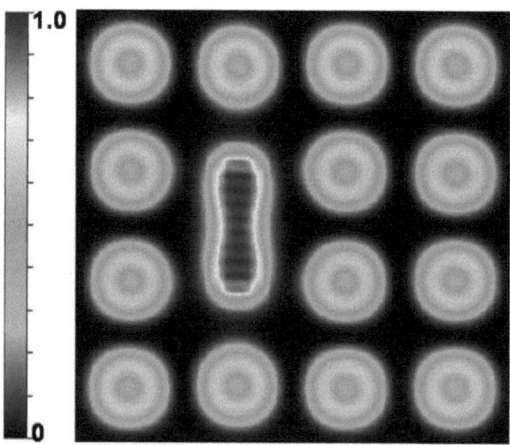

Figure 3.7: The ELF in the [100] plane shows that the two electrons of the M center are localized in the divacancy and form a quasi-bond leading to a self-stabilization effect in CaF_2.

3.2 Point defects in CaF$_2$

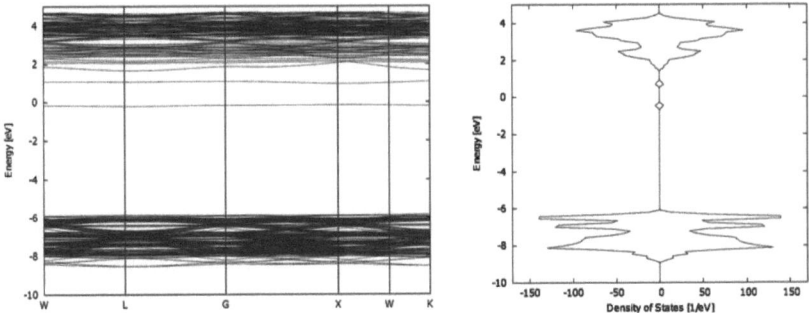

Figure 3.8: Band structure (left) and density of states (right) of the M center: The band structure shows well defined states in the bandgap with no dispersion.

at room temperature [101]. In my considerations regarding laser damage, it will therefore only play a minor role.

3.2.4 The H center

The H center is a F_2^- dimer like the V_k center, but is oriented along the (111) axis. One of the atoms resides close to a lattice site and the second one close to the interstitial site with a small displacement towards each other. This geometry was measured by Parker et al. [98] and is well confirmed in my calculations. The structure of the H center is shown in figure 3.9 and is easily identified in the ELF (figure 3.10). The calculated fully relaxed structure is shown in figure B.4).

The H center exhibits a state very low in the bandgap just above the conduction band (figure 3.11). Since the state is so close to the conducting band, the H center is not expected to be optically active, especially not in the DUV region. An optical transition of the H center is observed at 4.0 eV (310 nm) [98]. The ELF allows to recognize the structure of the H center, but it does not allow to see the localized hole. However, the localization of the hole can be observed in a spin density calculation (figure 3.10). The localization of the hole is not as strong as the localization of the F center, which allows the assumption that the diffusion barrier of the H center will be lower than that of the F center.

3.2.5 The self-trapped exciton

The self-trapped exciton (STE) in CaF$_2$ is formed by exciting an electron into the conduction band, thereby creating an electron-hole pair, which can localize via a lattice distortion forming an STE. Its structure is well-known as a nearest-neighbor F-H pair [102, 103, 104, 105, 106, 107].

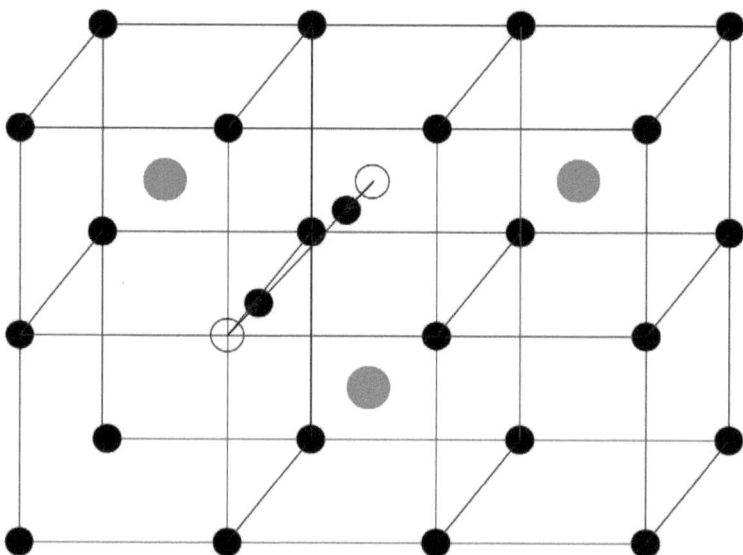

Figure 3.9: The structure of the H center: The two fluorine ions of the H center are located along the (111) axis connecting a lattice site and an interstitial site (empty circles). They are displaced towards each other forming an F_2^- dimer with the hole localized on the bond. The calculated fully relaxed structure is shown in figure B.4.

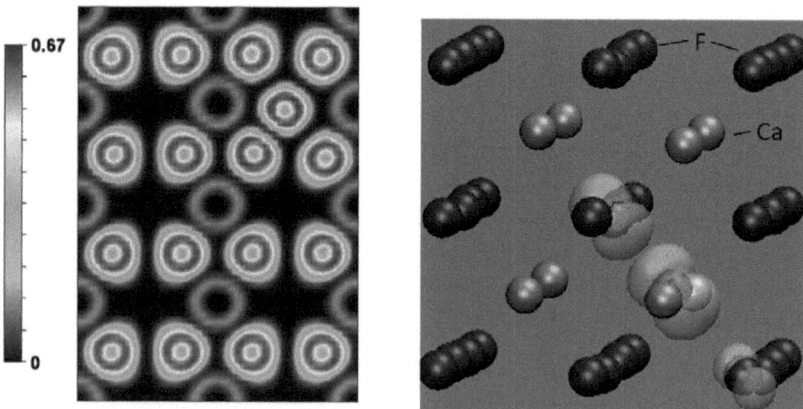

Figure 3.10: ELF (left) and spin density (right) of the H center: The ELF in the [110] plane shows the structure of the H center. The localization of the hole cannot be seen in the ELF. The spin density shows the localization of the hole on the H center.

3.2 Point defects in CaF_2

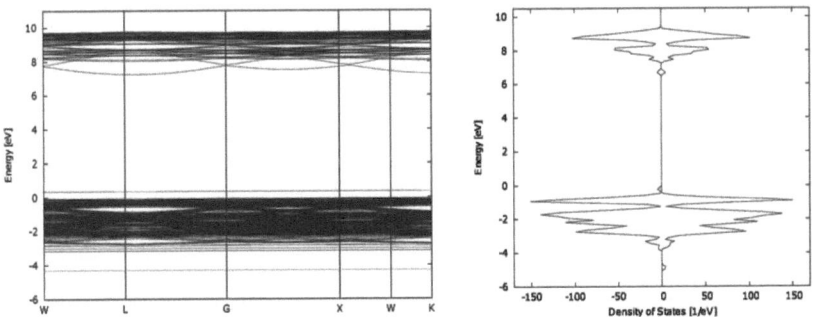

Figure 3.11: Band structure (left) and density of states (right) of the H center: The band structure shows a well defined occupied state in the bandgap close to the valence band.

The STE is of special interest in this work, because it is the initial defect structure formed in the lattice under DUV irradiation and is as such a precursor for stable F-H pairs, which subsequently also lead to other defects.

The formation mechanism of the STE in CaF_2 has been studied in detail [106, 107]. For the formation of an STE, first an electron-hole pair is formed by excitation of an electron into the conduction band. Then, in a first step the hole is localized as a V_k center, subsequently forming a $[V_k e]$ complex where the electron is bound by the V_k center. In the second step the V_k center undergoes a rotation and translation to the H center configuration and the electron localizes on the now vacant lattice site. The recombination of the STE can occur via an optically forbidden transition showing fluorescence at 278 nm with a lifetime of 1.1 μs at room temperature [108, 109].

F-H pair formation

The formation of stable F-H pairs, which are critical with respect to laser stability, occurs as the separation of the nearest-neighbor F-H pair, i.e. the STE. Since the STE itself as well as the F and H center distort the surrounding lattice, a finite attractive interaction between the F and the H center is expected on a very short length scale, because the distortion of the STE should be energetically favored over the separated distortions of the F and H center. This separation process therefore requires an activation energy in the range of the diffusion barrier of the F or the H center and is induced by the interaction of two excitonic events [106]. Therefore, the yield of stable, spatially separated F-H pairs is smaller than that of STEs by almost two orders of magnitude [106].

I have calculated the formation energy of an F-H pair by taking a perfect CaF_2 lattice and then forming an F and an H center which are infinitely far apart. Since both defects are

	formation energy [eV]
F-H pair	-7.52
F_{Na}-H pair	-1.34
F-H_Y pair	-0.73
M_{Na}-H pair	-7.33

Table 3.1: Calculated formation energy of the F-H pair in CaF_2. The impurity stabilized values assume that one of the defects is immediately stabilized by an impurity. For the M_{Na}-H pair the pair is formed at a F_{Na} center.

localized and charge neutral, this is a valid approach. The formation energy results from the abstract equation

$$CaF_2 \rightarrow F\ center + H\ center + \Delta E, \qquad (3.1)$$

which for the supercells used can be written as

$$2 \times Ca_{32}F_{64} \rightarrow Ca_{32}F_{63} + Ca_{32}F_{65} + \Delta E. \qquad (3.2)$$

The result for the formation energy is $\Delta E = -7.52\,eV$, which is in good agreement with the defect formation energies calculated by Franklin [110]. The negative value of the formation energy implies that energy needs to be put into the system to create an F-H pair. The formation energy of an F-H pair in CaF_2 and at impurities is shown in table 3.1.

3.3 Stabilized point defects in CaF_2

The stabilization of point defects is a crucial issue in laser damage processes. When the initial point defects stabilize in the material, i.e. they form stable defect structures, a permanent loss of the optical performance of the material is observed. Since the binding electrons in CaF_2 have been transferred from the Ca atoms to the F atoms, the conducting band is made up of the outer shell electrons of the fluorine. This means that defects in the fluorine structure will have an direct effect on the electronic structure of the material, while defects in the calcium structure only exhibit a direct effect via defect stabilization, unless the defect is an impurity with an optical band such as e.g. rare-earth ions. In the crystalline structure, it is the localization of electrons that leads to optically active color centers.

I have calculated the electronic properties for several stabilized defects. The results for the structures, band structures, and density of states for all defect structures are shown in appendix B. The energies of self-stabilization of point defects and their stabilization by impurities are shown in table 3.2.

3.3 Stabilized point defects in CaF_2

Stabilized defect	stabilization energy ΔE [eV]
self-stabilization	
$F + F \to M$	0.43
Stabilization by Na^+ impurity	
$F + Na \to F_{Na}$	6.17
$F + F_{Na} \to M_{Na}$	0.19
$H + Na \to H_{Na}$	0.05
Stabilization by Y impurity	
$H + Y \to H_Y$	6.78
$F + Y \to F_Y$	1.13
$F + F_Y \to M_Y$	0.30

Table 3.2: The energies for self-stabilization and stabilization by impurities of point defects.

3.3.1 Self-stabilization by M center formation

The formation of an M center can be described by the following equation

$$\text{F center} + \text{F center} \to \text{M center} + \Delta E, \quad (3.3)$$

where ΔE is the formation energy of the M center. A positive formation energy means that the final state is energetically favored with respect to the initial state. A negative formation energy implies, that energy needs to be put into the system to create the defects. The formation energy of the M center results to

$$2 \times Ca_{32}F_{63} \to Ca_{32}F_{64} + Ca_{32}F_{62} + 0.43\,eV. \quad (3.4)$$

For the H center a self-stabilization effect is not observed. Thus, the radiation induced F centers show self-stabilization, while the radiation induced H centers do not. This is an important finding on which I later base my laser damage model.

3.3.2 Stabilization by impurities

Impurities have a strong effect on the stabilization of point defects. Because of the high reactivity of fluorine, it is sufficient to take cation impurities into account. The only anionic defect discussed in literature is oxygen because of its abundance [26, 111, 112]. Since CaF_2 crystals are today grown in an oxygen free environment, anionic impurities are not taken into account.

	Ca^{2+}	Na^+	Y^{3+}
ionic radius [Å]	0.99	1.02	0.90

Table 3.3: The ionic radii of calcium, sodium, and yttrium.

Na^+ stabilized defects

As defect stabilizing impurities Na^+ and Y^{3+} are considered, because they have a similar ionic radius to Ca^{2+} (table 3.3) and are therefore easily incorporated in the CaF_2 crystal matrix.

Na^+ stabilized defects

Sodium is of special interest because it is incorporated into the CaF_2 matrix easily, and is present in abundance e.g. in form of salts in human sweat. Therefore, the production of sodium free CaF_2 is very demanding. Due to strong efforts in the past decade, CaF_2 can today be produced with an impurity content below the ppb range. However, during this time it was sodium that caused defect stabilization of radiation induced point defects and was identified as an enhancer of laser damage [14, 17]. The primarily investigated defect structures with sodium are the F_{Na} center and the M_{Na} center [36, 17], also referred to as the Na^+ stabilized F and M center respectively.

For laser damage purposes the F_{Na} center can be seen as a precursor, because the F_{Na} center itself is not optically active. The band structure of the F_{Na} center exhibits a state in the bandgap, but it is not occupied (figure B.5). In the F center this impurity is occupied by an electron, while in the F_{Na} center this vacancy is left empty because the monovalent Na^+ cation provides one electron less than the divalent Ca^{2+}, which leads to a stronger lattice distortion (compare figures B.2 and B.5). Since the bandgap state is not occupied and just below the conduction band, the excitation energy for electron-hole pair formation in the vicinity of an F_{Na} center is reduced, however the F_{Na} center is not optically active at 193 nm. The calculated formation energy for the F_{Na} center is 6.17 eV, which is in the range of the F-H pair formation energy. Therefore, the recombination of an F-H pair and the stabilization of the F center at a Na^+ impurity are competing processes. This implies that the formation of an F-H pair at a Na^+ impurity is energetically favored, due to the stabilization of the F center by the Na^+ impurity (table 3.1).

With the formation of an M_{Na} center, a second vacancy is stabilized at a Na^+ impurity forming a state with one localized electron in the divacancy. The band structure then exhibits a localized occupied state in the bandgap (figure B.6), which makes the M_{Na} center optically active at 193 nm. The optical properties of the M_{Na} center have been studied intensively [93, 36, 17], showing absorption bands at 380 nm (3.25 eV) and 615 nm (2.0 eV) as well as an emission band at 760 nm (1.6 eV). The formation energy of the M_{Na} center is calculated as the adsorption of a second F center to a F_{Na} center and results to 0.19 eV.

The stabilization of an H center by a Na^+ impurity is unlikely, because the H center is a localized hole and the monovalent Na^+ represents a hole in the lattice structure as well. The stabilization energy of the H center is almost zero, so that an effective stabilization is not taking place.

Y^{3+} stabilized defects

Yttrium impurities in CaF_2 are the trivalent counter part to Na^+ impurities as cationic impurities. The effect of Y^{3+} impurities on the optical properties [113] and F interstitial diffusion [114] has been investigated. From a laser damage perspective, doping with trivalent Y^{3+} is considered as a counterpart to the monovalent Na^+ [115].

An Y^{3+} impurity in CaF_2 leads to an occupied state in the conduction band (figure B.9). In an insulator this will not be a stable state, so that the stabilization of a defect is expected to have a high probability. Since Y^{3+} is trivalent, an Y^{3+} impurity should stabilize an H center because the extra electron provided by the Y^{3+} will compensate the hole of the H center. The F_2^- dimer of the H center results in two F^- ions, of which one is located on an interstitial site (figure B.8). This center does not have any occupied states in the bandgap and is thus optically inactive. The stabilization energy of the H_Y center is 6.78 eV, which is also in the range of the F-H pair formation. Therefore, also a Y^{3+} impurity stabilizes an F-H pair formed in its vicinity (table 3.1).

In addition to the H center, the Y^{3+} impurity can also stabilize an F center. The additional electron supplied by the Y^{3+} impurity is localized in the vacancy forming a doubly-occupied non-dispersed state in the bandgap (figure B.10). Therefore, the F_Y center is expected to be optically active. Its stabilization energy is 1.13 eV.

According to my calculations also a M_Y center is possible with three electrons localized in a divacancy. The band structure shows three occupied localized states in the bandgap (figure B.11). The stabilization energy results to 0.30 eV.

3.4 Diffusion properties of point defects

3.4.1 F-center diffusion properties

The diffusion barrier of the F center has been calculated in two different ways. For the first calculation, I calculated the diffusion barrier in the simplest possible manner, using the VASP code within the MedeA environment. Because of the high symmetry of CaF_2, I assumed that the diffusion path is straight. Then I simply calculated the energy of images along this path, which is the linear interpolation between the initial and final state, yielding $\Delta E_{F,VASP} = 1.50\,\text{eV}$. In a second step, I fixed the diffusing atom along the direction of diffusion and relaxed the surrounding structure. This lowered the barrier to $\Delta E_{F,VASP} = 0.90\,\text{eV}$. The results are shown in figure 3.12.

Pucchina *et al.* [116] state that the diffusion path is not straight because the diffusion barrier is lower if the diffusion path is deviated by 1.25 Å at the saddle point as the distance of the diffusing fluorine to the nearest two calcium atoms is thereby reduced. The suggested

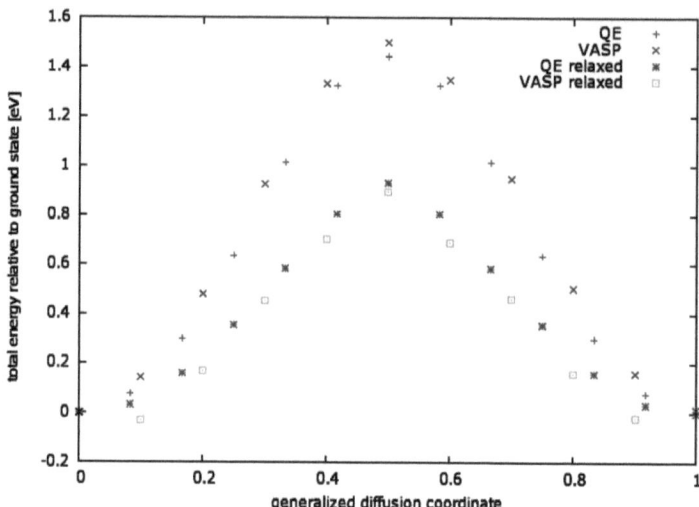

Figure 3.12: The diffusion barrier of the F center was calculated with VASP and Quantum Espresso. The upper two curves are the unrelaxed results, i.e. the interpolation between the initial and final state without any relaxation. The result from VASP (green) is $\Delta E = 1.50\,\text{eV}$, Quantum Espresso yields $\Delta E = 1.44\,\text{eV}$. The lower two curves show the barrier for the relaxed calculation. The result from Quantum Espresso was obtained by the nudged elastic band method and yields $\Delta E = 0.93\,\text{eV}$. The result from VASP ($\Delta E = 0.90\,\text{eV}$) was obtained by fixing the coordinate of the diffusing atom in the direction of diffusion and relaxing the surrounding lattice.

path deviation is depicted in figure 3.13. However, I could only confirm this assumption for the static, i.e. the non-relaxed case, which lowered the diffusion barrier by 0.18 eV.

In order to determine the true diffusion path of the F-center in CaF_2, I used the nudged elastic band method as described in chapter 2.3.2 as implemented in Quantum Espresso. Since the nudged elastic band method depends on the initial path guess, I performed calculations starting with the straight path, the displaced path as suggested by Pucchina et al., and a path displaced even further than that, as shown in figure 3.13. As a result for all starting paths, I obtained the straight path with a displacement of the nearest Ca atoms as the minimum energy path (figure 3.13). The series of images calculated for the final path are shown in figure 3.14. For comparison, I extracted the barrier of the initial path ($\Delta E_{F,\text{QE}} = 1.44\,\text{eV}$), which shows good agreement with the VASP result. For the minimum energy path, I obtained a barrier of $\Delta E_{F,\text{QE}} = 0.93\,\text{eV}$, also in excellent agreement with my previous result. The barriers are shown in figure 3.12.

3.4 Diffusion properties of point defects

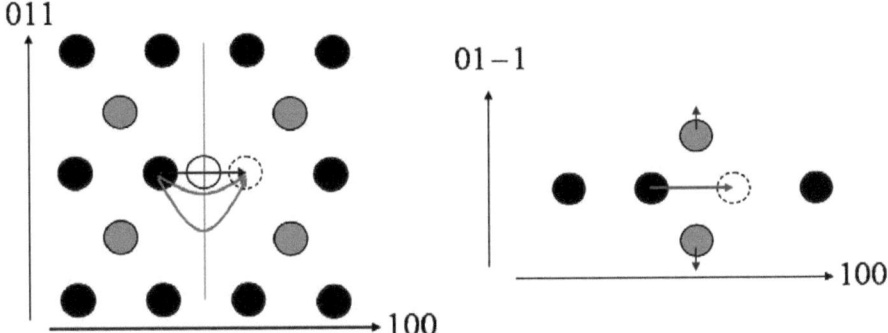

Figure 3.13: Left: The straight arrow shows the diffusion path of the F center. The slightly bent (middle) arrow indicates the diffusion path as proposed by Pucchina et al. [116]. The empty circle indicate the calcium ions above and below the diffusion plane The lower arrow indicates an exaggerated deviation as used as a starting path for the nudged elastic and method. Right: Rotating the system by 90° around the diffusion path shows that the fluorine atom is moving through the direct connection of two calcium ions. The nudged elastic band calculations show a displacement of the calcium atoms in the diffusion process rather than a deviation of the diffusion path.

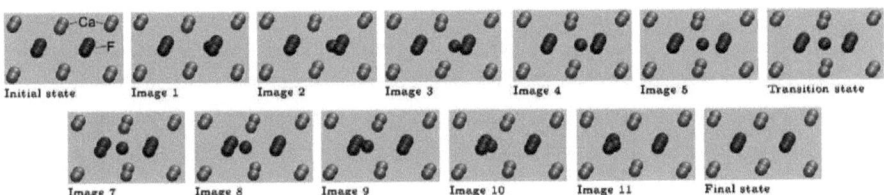

Figure 3.14: The images of the minimum energy path of the F center diffusion resulting from a nudged elastic band calculation. The diffusion path is straight.

Calculation of the prefactor

The first step to determine the prefactor is to calculate the phonon frequencies of the initial and transition state. They were calculated via linear response and with the frozen phonon method as described in chapter 2.3.3. Then the jumprate is calculated according to equation (2.34). Since the values of $h\nu/2k_BT$ are in the range of $0.13 - 1.21$, Vineyard's approximation of (2.34) results in an error of about 10%. The results for the jump rate are given in table 3.4.

The prefactor is then determined as described in chapter 2.3. For the F center the number of stable nearest neighbor sites is 6 and the probability for a jump to lead forward is 1/6 for the (100) direction. Therefore, the factor $n\beta$ in (2.32) is 1. $d = a/2$ is given by half of the lattice constant $a = 5.46$ Å. The results for the prefactor are shown in table 3.4.

F center	Γ_0 [THz]	$\Gamma_{0,\text{Vineyard}}$	D_0 [m²/s]	$D_{0,\text{Vineyard}}$
linear response	45.0	41.8	$3.4 \cdot 10^{-6}$	$3.1 \cdot 10^{-6}$
frozen phonon	52.2	50.8	$3.9 \cdot 10^{-6}$	$3.8 \cdot 10^{-6}$

Table 3.4: The jump rate Γ_0 and prefactor D_0 for the diffusion coefficient of the F center obtained from the calculation of the phonon modes according to (2.34) at $T = 300$ K without the exponential factor. The high temperature approximation by Vineyard (2.35) is given for comparison.

3.4.2 H-center diffusion properties

The calculation of the H center is more cumbersome than for the F center because it involves the movement of at least two atoms. The H center also has less symmetry and therefore allows more possible diffusion paths. The ground state configuration of the H center and the possible diffusion paths are shown in figure 3.15. Because the H center is not as strongly localized as the F center, I expect the diffusion barrier of the H center to be lower than the diffusion barrier of the F center.

The indicated sites 1-3 lie within the same fluorine cube as the initial H center. Site 1 is the displacement in (100) direction, site 2 in (110) direction, and site 3 in (111) direction. As the sites are all within the same fluorine cube, the diffusion paths are very short and I expect the barriers for these processes to be very low because they do not involve a strong distortion of the lattice. On the other hand these displacements do not lead to any "real" diffusion, because the H-center is only displaced within the fluorine cube. The results for the barrier are shown in figure 3.16 and confirm my expectations. However, a true diffusion requires the movement of both atoms of the H center.

Site 4 is the nearest stable site for the H center. Since the H center can only diffuse to a fluorine cube that does not contain a Ca atom, i.e. it is vacant and contains an interstitial site, it cannot diffuse to any directly adjacent cube, but only the diagonally adjacent cubes in (110) direction. I expect the diffusion path to be via the lattice site (along the respective arrow in figure 3.15), i.e. both atoms of the H center move. The atom at the lattice site moves towards the interstitial site (site 4) and the atom from the interstitial site moves to the position close to the lattice site.

The diffusion process to site 5 involves the movement of three atoms along the respective arrow in figure 3.15. In principle the final state of this path can also be reached by two diffusion hops equivalent to the hop to site 4. This means that this path is only relevant if the barrier is considerably lower than the barrier of the path to site 4. My results show that this is not the case (figure 3.16) from which I conclude that the relevant diffusion path for the H center is given by path 4. The result for the barrier of the minimum energy path of the H center diffusion is thus $\Delta E_H = 0.25$ eV.

3.4 Diffusion properties of point defects

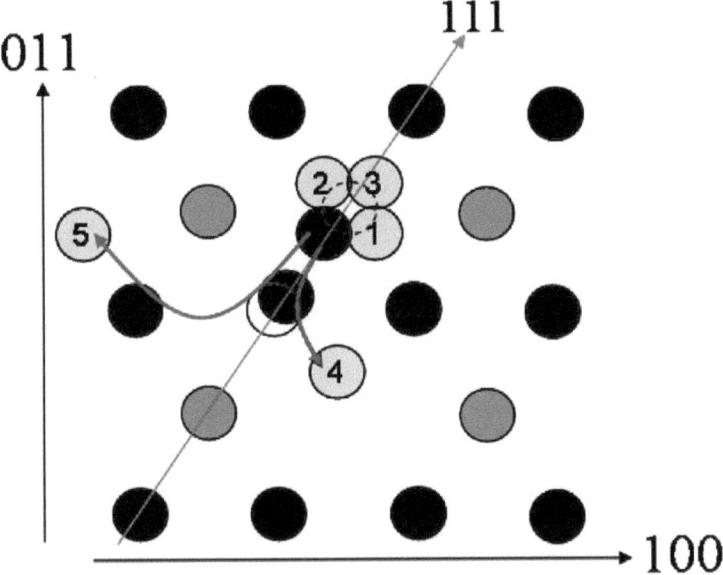

Figure 3.15: Configuration and possible diffusion paths of the H center. The empty solid circle indicates the fluorine lattice site, the dashed circle indicates the interstitial site. The two displaced black circles along the 111 axis show the equilibrium configuration of the H center. The sites 1-3 indicate three possible stable sites within the same (vacant) fluorine cube. Site 4 is the nearest stable site in the next vacant fluorine cube, the diffusion path is via the lattice site indicated by the respective arrow. Site 5 is in the next-to-nearest neighbor cell, as the nearest neighbor cell in 100 direction is occupied by a calcium. The diffusion path is along the fluorine cube edge indicated by the respective arrow.

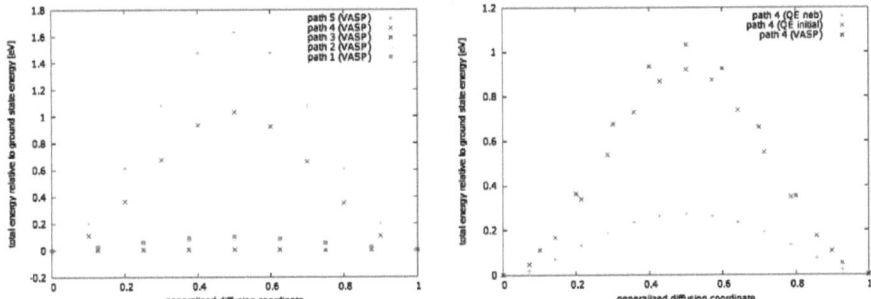

Figure 3.16: The diffusion barrier of the H center: First estimation of the diffusion barrier of the H center without any relaxation (left). The unrelaxed barrier for path 4 as well as the result from the nudged elastic band calculation (right).

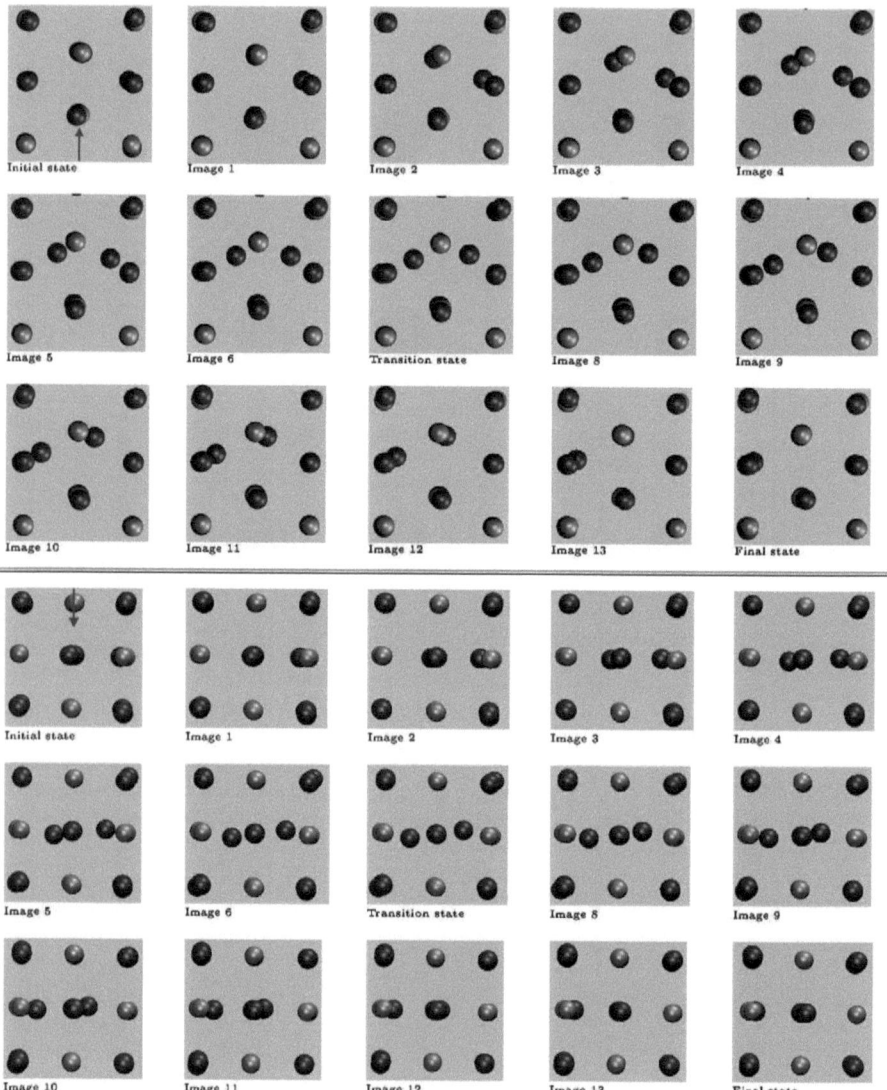

Figure 3.17: The images of the minimum energy path of the H center diffusion obtained from a nudged elastic band calculation. The arrow in the top picture indicates the viewing direction of the bottom pictures and vice versa. Top: The diffusion path in the [110] plane resembles path 4 as proposed in figure 3.15. Bottom: Viewed in the [100] plane the diffusion path of the two F atoms is deviated above and below the diffusion plane. This symmetry-breaking deviation keeps the distance between the two atoms constant.

3.4 Diffusion properties of point defects

The minimum energy path for the H center diffusion is shown in figure 3.17. The first set of images shows the diffusion path in the [100] plane as proposed. After a first nudged elastic band calculation the resulting path was perfectly in the plane shown in figure 3.15. However, two imaginary phonon modes indicated that this could not be the minimum energy path. By deviating the transition path as described in chapter 2.3.2, I obtained the minimum energy which exhibits a slight deviation of the diffusing atoms out of the symmetry plane. In this case the breaking of the symmetry allows the two F atoms of the H center to maintain the distance to one another, i.e. the F_2^- dimer maintains its bond length throughout the diffusion process.

In principle one could assume that the easiest diffusion of the H center would be along its axis, the (111) direction. However, in the fluorite structure every other fluorine cube is occupied by a cation, so that this diffusion path is not free and the resulting diffusion path is more complex.

Calculation of the prefactor

In principle the calculation of the phonon modes for the H center is analogous to those of the F center. However, we only calculated the phonon modes via the frozen phonon method because of limited computing time. This is justified as the results for the F center show good agreement. The results are shown in table 3.5. The H center diffusion along path 4 has stable final sites in the fluorine cubes which are connected to the initial cube at its edges. Since the cube has 12 edges and the diffusion process can take place via both lattice sites at the ends of each edge, there are 24 possible diffusion paths with 24 final stable sites. Considering diffusion in the (100) direction, the paths to 4 neighboring cubes, i.e. 8 paths, lead forward in the direction of diffusion, which means the probability for a diffusion jump in the forward direction is $\beta = 8/24 = 1/3$. The length of each jump projected on the diffusion direction is $d = a/2$, where I have assumed that the mean position of the interstitial fluorine of the H center is the interstitial site. This assumption is valid, because the diffusion barrier within the fluorine cube is so small, that the H center can easily associate with any of the fluorine atoms on the corners of the cube. The results for the prefactor are shown in table 3.5.

H center	Γ_0 [THz]	$\Gamma_{0,\text{Vineyard}}$	D_0 [m^2/s]	$D_{0,\text{Vineyard}}$
frozen phonon	1.0	1.3	$6.0 \cdot 10^{-9}$	$7.8 \cdot 10^{-9}$

Table 3.5: The jump rate Γ_0 and prefactor D_0 for the diffusion coefficient of the H center obtained from the calculation of the phonon modes according to (2.34) at $T = 300$ K without the exponential factor. The high temperature approximation by Vineyard (2.35) is given for comparison.

3.4.3 Dielectric measurements on CaF_2

Dielectric measurements on different CaF_2 samples from former Schott Lithotec have been performed at Schott's corporate research facility as well as the University of Augsburg [117] showing good agreement with one another. The results of one typical CaF_2 sample are shown in figure 3.18.

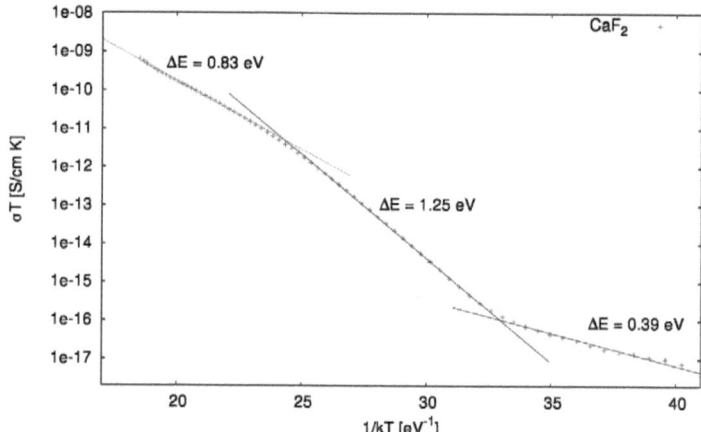

Figure 3.18: Conductivity of a CaF_2 sample. The dielectric measurements were performed by M. Aigner.

The current understanding of the two conducting mechanisms is the following. There is a very small but finite intrinsic concentration of F centers in CaF_2, while the intrinsic concentration of H centers is negligible [118]. Since the kinetic energy of the fluoride ions is Boltzmann-distributed, there is a small probability for F centers to have sufficient energy to overcome the diffusion barrier and move within the bulk crystal. This probability obviously increases with increasing temperature, which leads to an increase of the conductivity. At yet higher energies the Boltzmann-distribution allows the statistical creation of F-H pairs. Then, there are two different charge carriers, F centers and H centers. Since the H centers are more mobile than the F centers due to the lower diffusion barrier, the H centers will make the dominating contribution to the conductivity. Because of the extremely high purity of the sample I do not consider any impurity effects.

The behavior of the depicted curve can be described by Arrhenius law

$$\sigma \cdot T \sim exp\left(-\frac{\Delta E}{k_B T}\right), \tag{3.5}$$

and the slope can be identified with the height of the diffusion barrier of the charge carrier.

3.4 Diffusion properties of point defects

Therefore, I interpret the measurements as follows. The low temperature range on the right is neglected because the values are so small that they can be considered as static noise. In the middle region the conductivity is due to the diffusion of F centers. In the region from 25-23 eV^{-1} (200-230°C) the thermal formation of F-H pairs begins and the H centers start to dominate the conductivity. At higher temperatures the conductivity is then fully due to H center diffusion.

For the F center the experimentally measured diffusion barrier yields

$$\Delta E_F = 1.25 \pm 0.06 \, \text{eV}. \tag{3.6}$$

The measured diffusion barrier of the H center is

$$\Delta E_H = 0.83 \pm 0.01 \, \text{eV}. \tag{3.7}$$

I have only presented the experimental results for one CaF$_2$ sample here, which was chosen because it shows a characteristic conductivity.

3.4.4 Discussion of the results

Previous results

The diffusion of point defects in CaF$_2$ has been well investigated experimentally. As early as 1956 Ure [119] determined the diffusion barrier of the F and H center. Since then a series of experimental [38, 114, 120, 124, 125, 100, 121] and a few theoretical [123, 122, 116] studies on the diffusion properties of defects in CaF$_2$ have been performed. The previous results are summarized in table 3.6.

The experimental values for the F center diffusion are in good agreement with each other showing a barrier height of $\Delta E_F = 0.52 - 0.86$ eV. The measured values for the H center on the other hand scatter over a wide range ($\Delta E_H = 0.46 - 1.6$ eV).

It is important to point out that the V_k centers can only be observed in measurements performed while the crystal is irradiated [100, 121] or if the crystal is cooled down to liquid nitrogen temperatures as the V_k center is not stable at room temperature [101, 34]. If the material is not irradiated during measurement, the V_k center will have recombined with an electron or will have formed a self-trapped exciton at the time of measurement.

All previous measurements were performed before 2000. However, in the past 10 years the impurity concentrations in CaF$_2$ have been reduced drastically to below the ppb range due to the high requirements of the semiconductor industry. Therefore, I assume that previous results could have been influenced by impurity contamination, which could lead to a lower diffusion barrier for the F center and a broader range for the diffusion barrier of the H center.

	ΔE_F [eV]	ΔE_H [eV]	
experimental			
Ure [119]	$0.52 - 0.86^a$	1.6^b	
Atobe [100]	0.72	0.46	0.33 (V_k-center)^c
Keig, Coble [120]	0.54		
Southgate [114]		0.9	
Huisinga et al. [121]			0.30 (V_k-center)^d
theoretical			
Pucchina et al. [116]	1.69		path deviation: 1.25 Å
Keeton, Wilson [122]	$0.3 - 0.4$	$0.6 - 1.5$	semiclassical potentials
Chakravorty [123]	0.64	1.56	Born model
present work			
VASP, relaxed	0.90		
QE, neb	0.93	0.27	
VASP, unrelaxed	1.50	1.03	
QE, unrelaxed	1.44	0.92	
experiment	1.25	0.83	

^{a} $200 - 690\,°C$
^{b} $690 - 920\,°C$
^{c} thermoluminescence measurement
^{d} photoconductivity measurement

Table 3.6: Numerical values for the diffusion barriers of the F-center and in the H-center. The observation of the V_k center diffusion is only possible, while irradiation of the material.

Previous theoretical results for the diffusion barrier have been obtained from semiclassical calculations [122], Born model calculations [123], and *ab inito* Hartree-Fock calculations [116].

My results

My calculations and the measurements show a good consistency with one another (table 3.6). The fully relaxed barrier for the F center is at 0.9 eV and the unrelaxed barrier lies at 1.5 eV. The measured value of 1.25 eV lies in between. For the H center the unrelaxed barrier is $0.92 - 1.03\,\text{eV}$, the relaxed barrier is 0.27 eV, and the measured barrier is again in between these values at 0.83 eV.

The fully relaxed barrier resembles a diffusion process, which is so slow compared to the lattice dynamics, that the lattice will fully relax throughout the whole diffusion process. This can be imagined as an infinitely slow diffusion process. The unrelaxed barrier on the other hand resembles a diffusion process, which is much faster than the lattice dynamics, so that the lattice will not relax at all. This diffusion process can be imagined as infinitely fast or adiabatic. The diffusion process in a real material will lie in between these two extremes.

To determine whether the diffusion process is slower or faster than the lattice dynamics,

3.4 Diffusion properties of point defects

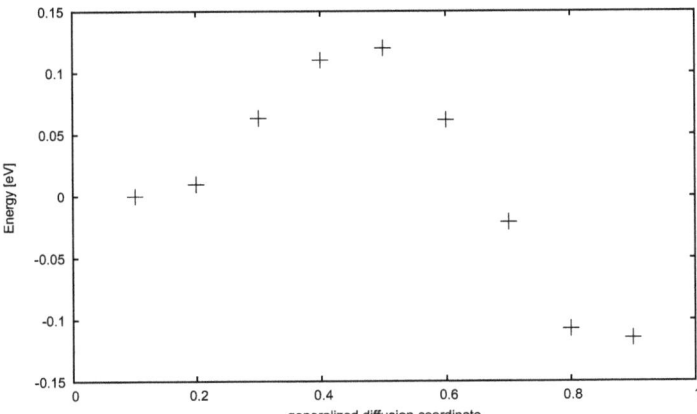

Figure 3.19: The diffusion barrier of F_{Na} center formation: A relaxation of the images was necessary in the calculation because of the stronger lattice distortion around the impurity. The Na^+ impurity lowers the barrier significantly. The resulting barrier is 0.12 eV in forward direction and 0.39 eV in the backwards direction.

I estimate the relaxation time of the lattice. An upper bound for the velocity of the lattice atoms is given by the Debye temperature, which is equivalent to the energy of the highest available phonon mode. The Debye temperature for CaF_2 is $k_B T_{Debye} = 43.9\,\text{meV}$ [126]. If the energy of the diffusing atom at the saddle point is much higher than the Debye temperature, the diffusion process is much faster than the lattice relaxation, and the surrounding lattice does not relax during the diffusion process, which implies that the barrier should be closer to the unrelaxed barrier. If the diffusion energy at the saddle point is much less than the Debye energy, the lattice has time to relax, and the barrier is expected to be closer to the fully relaxed calculation. Since the diffusion barriers in CaF_2 are far above the Debye temperature, I expect the experimental barriers to be closer to the unrelaxed barriers.

The kinetic energy of the atoms in CaF_2 is Boltzmann distributed. Since the room temperature energy of $k_B T = 25\,\text{meV}$ is very low compared to the height of the diffusion barriers, diffusion processes will only take place for atoms in the high-energy tail of the Boltzmann-distribution. It follows directly that the diffusion process for the H center is more likely than for the F center. In addition, I conclude that the kinetic energy at the saddle point for the diffusing atoms in the H center is higher than for the F center. The difference in kinetic energy should be approximately the difference of the barrier heights. Therefore, I expect the measured value for the H center to be closer to the unrelaxed barrier than for the F center, which is in good agreement with my calculations and the measurements.

In comparison with previous results my barriers are higher. Most likely this is due to the fact that in recent years the impurity content of synthetically grown CaF_2 crystals has been lowered significantly. Since an increased impurity content lowers the diffusion barrier of the F centers considerably, it is very likely that previous measurements have been influenced by a rather high impurity content. In order to verify that impurities actually lower the diffusion barrier of the F centers, I have calculated the barrier for an F center forming an F_{Na} center (figure 3.19). Even though the calculation is only a rough estimate, the resulting diffusion barrier is lowered significantly.

For the H center my results are in the lower range of previous measurements. The result for the minimum energy path is lower than previous measurements, but in accordance with my assumptions mentioned above. Southgate [114] observed that the diffusion barrier of fluorine interstitials in Y^{3+} doped CaF_2 is 0.3 eV higher than in pure CaF_2, which would imply that earlier measurements overestimated the real barrier height.

The measurements by Atobe [100] confirm my assumption that the F center barrier is larger than the H center barrier and that the V_k barrier is even lower. I conclude that the H centers in CaF_2 are more mobile than the F centers, which is an important result, that I will later incorporate in my model of laser damage in CaF_2.

3.5 Optical properties of Ca colloids in CaF_2

3.5.1 Lambert-Beer law

The attenuation of light in a material is described by the Lambert-Beer law. The transmittance T is given by

$$T = \frac{I}{I_0} = e^{-\alpha d}, \qquad (3.8)$$

where I is the intensity, I_0 the initial intensity, α the attenuation coefficient, and d the thickness of the material. Optical materials often exhibit extinction in addition to the intrinsic attenuation, which is caused by scattering or absorption centers, such as color centers, impurities or also colloids. I will refer to the intrinsic absorption of CaF_2 as attenuation, while any additional loss of radiation will be referred to as extinction, which is the sum of absorption and scattering by any defects or impurities. The Lambert-Beer law can then be written as

$$T = \frac{I}{I_0} = e^{-(\alpha+k)d}, \qquad (3.9)$$

where α is the intrinsic attenuation coefficient of CaF_2 and k is the turbidity coefficient related to any defect structures or impurities. Since the turbidity describes absorption and especially scattering in a material, it is often also referred to as haze.

The turbidity coefficient can be calculated from the extinction cross section σ_{ext} of the

3.5 Optical properties of Ca colloids in CaF$_2$

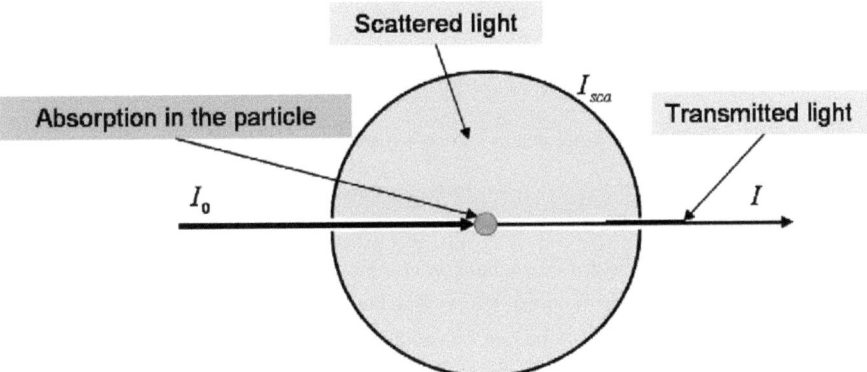

Figure 3.20: Schematic illustration of extinction, absorption and scattering: The energy of the absorbed radiation stays in the particle. The scattered radiation is emitted in a different direction than the incident direction. The extinction is the sum of the two. The transmitted light is the fraction that is not absorbed or scattered.

scattering and absorbing defect structure

$$k = k_{ext} = c\sigma_{ext}, \qquad (3.10)$$

where c is the number of particles per unit volume and k_{ext} is the extinction coefficient. The extinction cross section is given as the sum of the absorption cross section σ_{abs} and the scattering cross section σ_{sca}

$$\sigma_{ext} = \sigma_{abs} + \sigma_{sca}, \qquad (3.11)$$

as long as multiple scattering does not occur, which is the case if $c\sigma_{sca}d \ll 1$ and can be assumed for CaF$_2$. Thus, the turbidity coefficient can be expressed as

$$k_{ext} = k_{abs} + k_{sca}, \qquad (3.12)$$

where $k_{abs} = c\sigma_{abs}$ and $k_{sca} = c\sigma_{sca}$ are the absorption coefficient and scattering coefficient respectively. For clarity, the relation of scattering, absorption, extinction, and transmittance are shown in figure 3.20.

For the investigation of laser damage, the intrinsic attenuation is only important for the initial absorption of light. Laser induced defects will not change the intrinsic attenuation, and therefore only the change of the extinction after a certain radiation period is of interest. The general Lambert-Beer law can be rewritten in the form

$$\alpha = -\frac{\ln(T)}{d} \tag{3.13}$$

and thus the so-called induced extinction can be expressed as

$$k_{\text{ind ext}} = \frac{-\ln T_{\text{before}} + \ln T_{\text{after}}}{d}. \tag{3.14}$$

In this form all intrinsic absorption of light is eliminated and a measure for the change of transmittance due to radiation induced defects is defined.

For metallic colloids the extinction can be calculated from

$$k_{\text{ext}} = \frac{\eta}{V_{\text{colloid}}} \sigma_{\text{ext}}, \tag{3.15}$$

where η is the colloid fractional volume, i.e. the fraction of the bulk volume that is occupied by colloids, V_{colloid} is the volume of the single colloids, and it is assumed that the colloids are monodisperse. Absorption and scattering are calculated analogously. The extinction, absorption and scattering cross sections are calculated from Mie-theory in chapter 3.5.3.

3.5.2 Optical properties of CaF_2 and Ca

The optical properties of a material are described by the complex refractive index $N = n + ik$ or the complex dielectric function $\epsilon = \epsilon' + i\epsilon''$. For a non-magnetic material the relation between the two is given by

$$\epsilon' = n^2 - k^2 \tag{3.16}$$

$$\epsilon'' = 2nk \tag{3.17}$$

$$n = \sqrt{\frac{\sqrt{\epsilon'^2 + \epsilon''^2} + \epsilon'}{2}} \tag{3.18}$$

$$k = \sqrt{\frac{\sqrt{\epsilon'^2 + \epsilon''^2} - \epsilon'}{2}}. \tag{3.19}$$

Refractive index of CaF_2

For a transparent material the imaginary part of the refractive index is zero. The wavelength dependent refractive index for CaF_2 is taken from [6] and is given by the Sellmeier formula

$$n^2 - 1 = \frac{B_1 \lambda^2}{\lambda^2 - C_1} + \frac{B_2 \lambda^2}{\lambda^2 - C_2} + \frac{B_3 \lambda^2}{\lambda^2 - C_3}, \tag{3.20}$$

3.5 Optical properties of Ca colloids in CaF$_2$

where the wavelength λ is given in μm. The Sellmeier coefficients for CaF$_2$ are given in table 3.7.

B_1	$6.188140 \cdot 10^{-1}$	C_1	$2.759866 \cdot 10^{-3} \, \mu\text{m}^2$
B_2	$4.198973 \cdot 10^{-1}$	C_2	$1.061251 \cdot 10^{-2} \, \mu\text{m}^2$
B_3	3.426999	C_3	$1.068123 \cdot 10^{3} \, \mu\text{m}^2$

Table 3.7: The Sellmeier coefficients for CaF$_2$ valid for 184 nm $< \lambda <$ 2326 nm [6].

Refractive index of Ca

The wavelength dependent complex refractive index for metallic calcium was taken from [127] and [128], who measured the refractive index in different wavelength regions. The experimental values were fitted with the ansatz

$$\epsilon = \epsilon_0 + \sum_j \frac{\omega_{pj}^2}{\omega_j^2 - \omega^2 - i\gamma_j \omega}, \tag{3.21}$$

of the multiple-oscillator model, where for the jth oscillator ω_{pj} is the plasmon frequency, ω_j is the resonant frequency, and γ_j is the damping factor. This ansatz allows to fit the experimental data and ensures that the Kramers-Kronig relations are fulfilled at the same time. The complex refractive index of Ca is shown in figure 3.21. For the calculation of cross sections from Mie scattering, a correction of the refractive index of Ca for small colloids is taken into account. This correction is necessary because for small particles the mean free path of the electrons is limited by the particle size. Additionally, a small correction due to the pressure exerted by the surrounding matrix is made. These corrections are discussed in appendix A.

3.5.3 Cross sections from Mie theory

In 1908 Gustav Mie developed the basic equations for the scattering of light on a spherical particle with a complex refractive index in a surrounding transparent matrix [21]. The basic problem is described by an incident plane wave on a spherical particle. The basic approach and a derivation of the solutions for the scattering cross sections are given in appendix A. The cross sections are functions of the incident wavelength and the particle size with the refractive indices of the spherical particle and the surrounding matrix as parameters.

Extinction, absorption and scattering at different colloid sizes

The results for theoretical extinction, absorption and scattering spectra of metallic Ca colloids in CaF$_2$ are shown in figure 3.22. The colloids show two characteristic peaks, one at 200 nm and one between 520–650 nm. The position of the first extinction peak at 200 nm does not

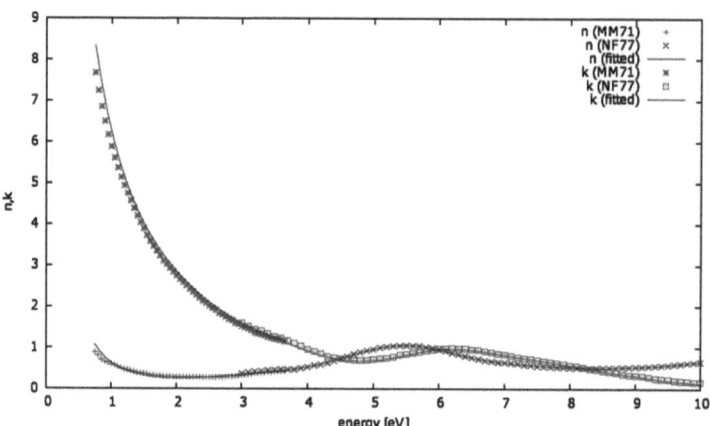

Figure 3.21: The real and imaginary part of the refractive index of metallic Ca. The data was taken from Mathewson and Myers (MM71) [127] and Nilsson and Forssell (NF77) [128]. The Kramers-Kronig relations were used for the fit.

change with increasing colloid size. Its intensity decreases between 10 and 50 nm before almost vanishing at larger colloid sizes.

The second peak is centered at 520 nm for small colloids up to a few nm in radius. This is in good agreement with the absorption band of the M center at 520 nm. Since the colloids grow as an agglomeration of F centers in CaF_2, the formation of the M center can be seen as the first step towards colloid formation. The fact that the absorption bands of the M center and very small colloids coincide shows that the two defect structures are related. While the formation of an M center from two F centers shifts the absorption band of the F center from 378 nm to the M center absorption at 520 nm, further agglomeration of F centers does not seem to have a strong effect on the absorption band. With increasing colloid size the peak shifts to about 580 nm at a colloid radius of 25 nm, for a colloid radius of 50 nm the peak position has shifted to 650 nm.

Further it is important to observe that the contribution of absorption and scattering changes with the colloid size. For very small colloids scattering effects are negligible. In the range of 22 to 50 nm the ratio of absorption to scattering changes from predominant absorption for small colloids to predominant scattering for large colloids. This can be explained with the penetration depth of light into the metallic colloids.

3.5 Optical properties of Ca colloids in CaF$_2$

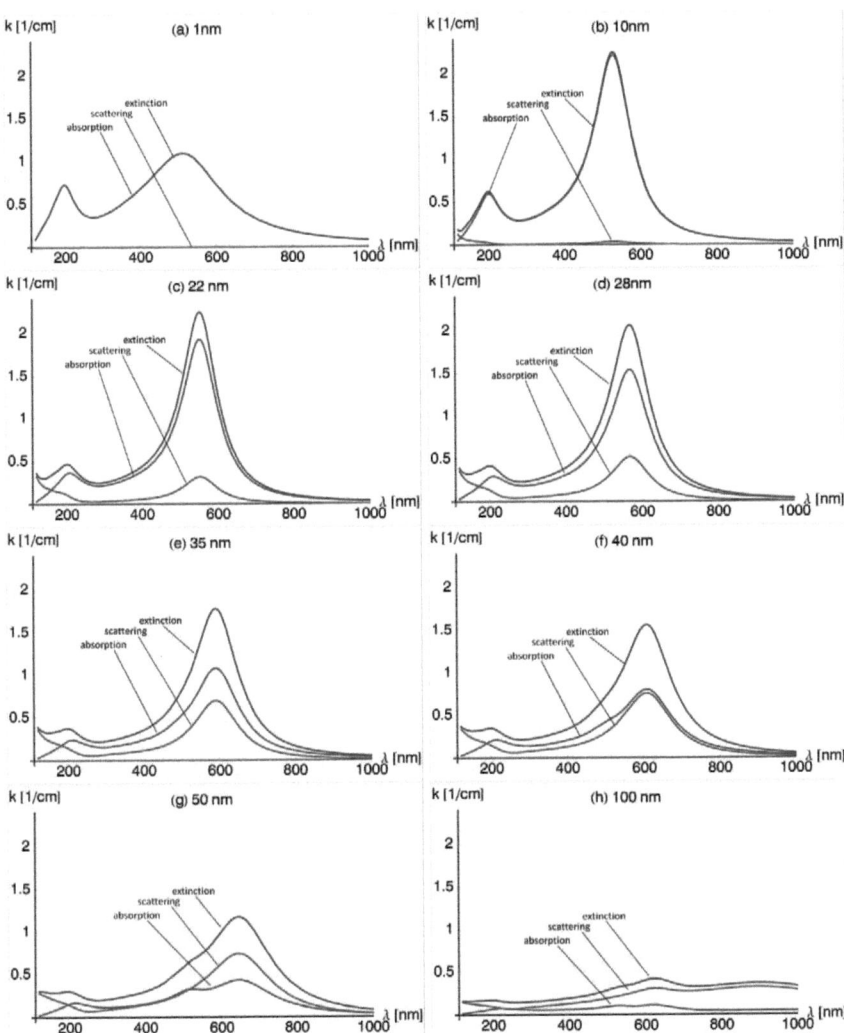

Figure 3.22: Extinction, absorption, and scattering of Ca colloids in CaF$_2$ calculated from Mie-theory for different colloid radii. The colloidal fractional volume was assumed to be 10^{-6}. For small colloids the extinction is dominated by absorption, larger colloids show stronger scattering and the total extinction decreases.

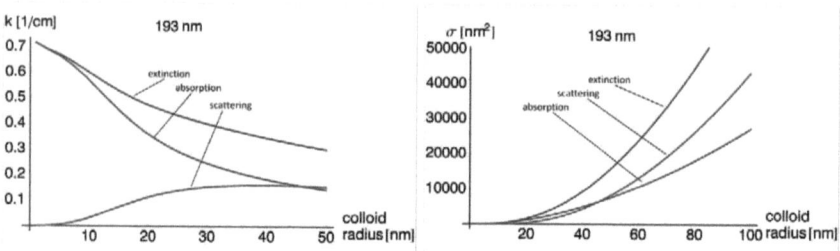

Figure 3.23: Left: Extinction, absorption, and scattering coefficients at 193 nm as a function of the colloid radius. The fractional colloid volume used is 10^{-6}. Right: Extinction, absorption, and scattering cross sections for one Ca colloid at 193 nm as a function of the colloid radius.

Extinction at a fixed wavelength

Extinction, absorption, and scattering coefficients and the corresponding cross sections of Ca colloids in CaF_2 at 193 nm are shown in figure 3.23. For colloid radii above 20 nm the increase of the extinction cross section with the colloid size is approximately quadratic, which means that the cross section is proportional to the geometric cross section in this region.

Energy absorption of colloids

It is straight forward to calculate the energy absorbed by a colloid during radiation from the absorption cross section. The energy absorbed by a colloid during a laser pulse with a certain fluence H_0 is

$$E_{\text{abs}} = \sigma_{\text{abs}} H_0. \tag{3.22}$$

For colloids with radii between 1 and 5 nm and fluences between 10 and 120 mJ/cm² the absorbed energy lies between $1.9 \cdot 10^2$ eV and $2.6 \cdot 10^5$ eV.

The extinction itself is scaled with the number of colloids per unit volume, where a constant fractional colloid volume is assumed. Very small and presumably well dispersed colloids show a stronger absorption than fewer larger colloids. This will be an important fact later, because the absorption at small sizes influences the growth of the colloids.

3.6 Formation energy of Ca colloids

In addition to the optical properties, I also calculate the formation energy of Ca colloids in CaF_2. The formation of a metallic colloid is the agglomeration of many F centers, i.e. fluorine vacancies. The main contribution to such a colloid is thus the "binding" energy of an F center to a colloid. In addition, contributions of the formation of a Ca/CaF_2 interface and the mechanical stress of the colloids need to be considered.

3.6 Formation energy of Ca colloids

	c_{11} [GPa]	c_{12} [GPa]	s_{11} [TPa^{-1}]	s_{12} [TPa^{-1}]
CaF$_2$	165	46	6.94	-1.53
Ca	16	8	94	-31

Table 3.8: The elastic stiffness constants c_{ij} and the elastic compliance constants s_{ij} of Ca and CaF$_2$ [129].

3.6.1 Binding energy of an F center to a colloid

I estimate this binding energy as the average binding energy of 64 F centers forming a colloid,

$$64 \times \mathrm{Ca}_{32}\mathrm{F}_{63} \rightarrow \mathrm{Ca}_{32} + 63 \times \mathrm{Ca}_{32}\mathrm{F}_{64} + 64\, E_{\mathrm{bind}}, \qquad (3.23)$$

yielding $E_{\mathrm{bind}} = 0.586\,\mathrm{eV}$ per F center. The energy is in the same range as the formation energy of the M center, which was to be expected, because the formation of an M center is the first step in the agglomeration of F centers.

3.6.2 Energy of the Ca/CaF$_2$ interface

The contributions of the colloid surface to the formation energy can be broken down into two components. First, the chemical energy of the surface, i.e. an adhesion energy, and second the relaxation of the surface due to the mechanical stress which is induced by the CaF$_2$ matrix. Because the stiffness of CaF$_2$ is about one order of magnitude higher than that of Ca (table 3.8), I assume that the mechanical stress is taken up by the metallic colloid completely and therefore neglect any surface relaxation effects. This implies that all calculations are performed at the lattice constant of CaF$_2$ without structural optimization. This is possible because the lattice structure of the Ca atoms in metallic Ca and CaF$_2$ is the same. The small change of the lattice constant is considered separately as mechanical stress.

To determine the interface energy from a quantum mechanical calculation, a supercell is constructed by stacking n CaF$_2$ unit cells and m Ca unit cells on top of each other forming a $(\mathrm{CaF}_2)_{n/2}\,(\mathrm{Ca})_m\,(\mathrm{CaF}_2)_{n/2}$ stack with periodic boundary conditions. This results in a stack of Ca and CaF$_2$ layers, which extend infinitely in two dimensions with a periodic layered structure in the third dimension. The interface energy is then calculated from the following abstract reaction

$$(\mathrm{Ca})_m + (\mathrm{CaF}_2)_n \rightarrow (\mathrm{CaF}_2)_{n/2}\,(\mathrm{Ca})_m\,(\mathrm{CaF}_2)_{n/2} + \Delta E, \qquad (3.24)$$

which can also be depicted symbolically

■ + ▢ = ▢■▢ + ΔE

where the dark shade symbolizes Ca and the light shade symbolizes CaF$_2$. The Ca cells were placed in the middle of the stack for symmetry reasons.

In order to avoid self-interaction of the layers, both layers need to have sufficient thickness. Therefore, I start with a Ca layer thickness of one lattice constant and calculate the adhesion energy for increasing CaF$_2$ layer thickness. Then, the number of Ca layers is successively increased until the values for the adhesion energy converge. Because the calculation of large stacks is very time consuming, the energy is first converged with respect to the Ca layer thickness, which is shown in figure 3.24. A layer thickness of two lattice constants was sufficient for the Ca layers.

Because the stiffness constants of CaF$_2$ are about 10 times larger than those of Ca (table 3.8), the necessary CaF$_2$ layer thickness is expected to be roughly 10 times the layer thickness of the Ca layer. Unfortunately, the necessary thickness of the CaF$_2$ layers could not be reached due to limited memory in the available computing machines. These calculations were attempted for the [100] interface and the [111] interface (figure 3.25). However, the interface energy shifts toward smaller absolute values, which means that an upper boundary for the absolute interface energy of $|E_{\text{surf}}| < 3.6\,\text{eV}/\text{nm}^2$ can be given.

3.6.3 Mechanical stress in colloids

Mechanical stress in Ca colloids is induced by the surrounding CaF$_2$ matrix because the lattice constant of CaF$_2$ is 2% smaller than the lattice constant of metallic Ca (table 3.9). As the stiffness of metallic Ca is much lower than that of CaF$_2$, I assume that the complete stress is absorbed by the metallic Ca.

In general the relation between the stress tensor σ_{ij} and the deformation tensor ϵ_{ij} is given by

$$\sigma_{ij} = c_{ijkl}\epsilon_{kl} \quad \text{or} \quad \epsilon_{ij} = s_{ijkl}\sigma_{kl} \tag{3.25}$$

where c_{ijkl}, the elements of a fourth rank tensor, are the elastic stiffness constants and s_{ijkl} are the elastic compliance constants. Because the three dimensional fourth rank tensors c_{ijkl} and s_{ijkl} are symmetric in ij and kl it is customary to reduce them to six dimensional second rank tensors c_{ij} and s_{ij} within the so-called matrix notation [130]. The stiffness and elastic compliance constants for CaF$_2$ and Ca are given in table 3.8.

	lattice constant [Å]
CaF$_2$	5.46
Ca	5.58

Table 3.9: The lattice constants of Ca and CaF$_2$.

3.6 Formation energy of Ca colloids

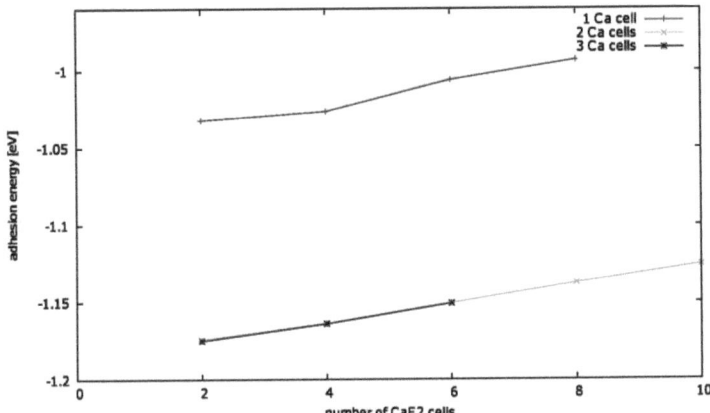

Figure 3.24: The surface energy of the Ca/CaF$_2$ interface plotted against the number of CaF$_2$ cells per layer. Already for a stack thickness of 2 Ca cells the interface energy is converged with respect to the number of Ca unit cells per stack.

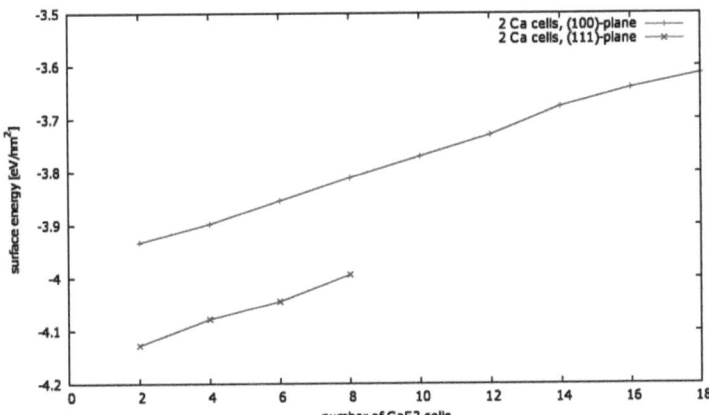

Figure 3.25: The surface energy of the Ca/CaF$_2$ interface for the [100] and [111] orientation as a function of the number of CaF$_2$ cells per stack. Larger stacks could not be calculated due to memory limitations. For the [111] direction numerical instabilities occurred for larger stacks.

For an isotropic deformation of a cubic material the relation reduces to

$$\sigma_i = \frac{1}{\kappa}\epsilon_i = B\epsilon_i, \qquad (3.26)$$

where $\kappa = 3(s_{11} + 2s_{12})$ is the compressibility and B is the bulk modulus. The stress energy per unit volume is given by

$$E_{\text{stress}} = \int \sigma_i \, d\epsilon_i \qquad (3.27)$$

$$= 3\int B\epsilon \, d\epsilon \qquad (3.28)$$

$$= \frac{3}{2}B\epsilon^2, \qquad (3.29)$$

where the deformation $\epsilon = \Delta l/l$ for a Ca colloid in CaF$_2$ is given by the relative change of the lattice constant

$$\epsilon = \frac{\Delta l}{l} = \frac{a_{\text{Ca}} - a_{\text{CaF}_2}}{a_{\text{CaF}_2}}. \qquad (3.30)$$

The resulting stress energy per unit volume in Ca colloids in a CaF$_2$ is thus $E_{\text{stress}} = 0.06\,\text{GPa} = 0.39\,\text{eV/nm}^3$.

The energy contribution from mechanical stress in the colloids is much smaller than the contributions of the F center binding and the surface energy, which is in good agreement with the assumption that relaxation energies can be neglected as they would be of the same order of magnitude or smaller.

3.6.4 Formation energy

The contributions to the formation energy of a Ca colloid in CaF$_2$ are listed in table 3.10. The formation energy of a Ca colloid and the single contributions are shown in figure 3.26.

For large colloid sizes the volume dependent binding energy dominates the formation energy of a colloid, which means that energy is gained by the successive agglomeration of F centers, and colloids could in principle grow infinitely large. However, this is an idealized picture, because it assumes that there is an infinite amount of F centers available and that there are no H centers for recombination. In a real material this is of course not the case. First of all, the number of available F centers is limited by the amount of radiation-induced F centers, and for every

Formation energy in	[eV/F center]	[eV/nm^3]
E_{bind}	0.59	28.80
E_{stress}	$-7.95 \cdot 10^{-3}$	-0.39
E_{surf}		$0 > -3.6\,\text{eV/nm}^2$

Table 3.10: Contributions to the formation energy of a Ca colloid in CaF$_2$.

3.6 Formation energy of Ca colloids

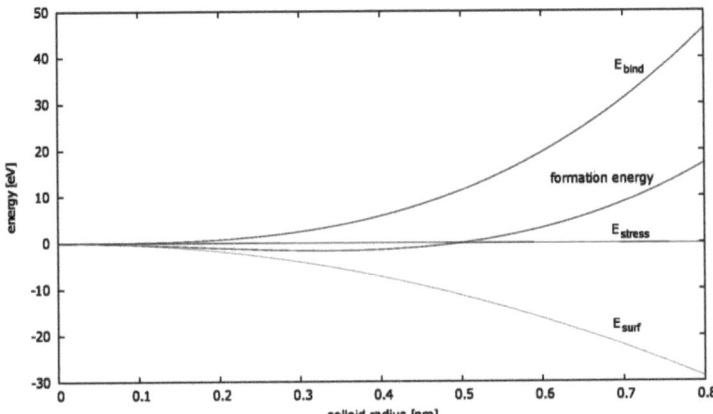

Figure 3.26: The formation energy of a Ca colloid in CaF$_2$ and the contributions from the binding of the F centers, the mechanical stress, and the surface energy.

radiation induced F center there is also an H center created in the material, which also allows for recombination of F and H centers. I will later discuss that the F center concentration can be higher than the H center concentration and that thus colloids are formed, but the conditions for unlimited growth are not provided in the real material.

Further the calculations show that up to a colloid radius of 0.5 nm, the formation energy is negative, i.e. the system only gains energy from the growth of colloids once they are larger than 0.5 nm. In an irradiated crystal sufficient energy is supplied by the radiation to initially form colloids. After having surpassed the minimum threshold, the colloids can grow by accumulating more F centers. At the same time small colloids absorb very strongly, which means that they can also be decomposed in the irradiation process. In addition, the colloids could grow in a process similar to Ostwald-ripening [131], where large colloids grow while smaller colloids shrink and dissolve. By this process the total number of colloids is reduced while the size of the colloids increases, which leads to a minimization of the surface of the colloids.

Chapter 4

Phenomenology of laser damage in CaF$_2$

Semiconductors for microchips are structured with optical microlithography. The semiconductor wafer is first coated with a photoresist, and then the desired structure is imaged onto the wafer. The actual structuring is then etched into the wafer, and finally the photoresist is removed leaving the structured wafer.

A central issue in the construction of a microchip is heat dissipation because it needs to be assured that the heat generated by the currents on the microchip can be transported and cooled properly. The heat development on a chip is determined by the flowing current, which is determined by the clock rate and the amount of electrons that need to be transported. The necessary electron movement can be reduced by making the structures on the microchips smaller, which then in return allows for higher clock rates making processors faster. Therefore the semiconductor industry has been striving towards the manufacturing of smaller and smaller structures on microchips following Moore's law [9]. One of the limiting issues is the optical resolution of the microlithographic structuring of the wafers.

The angular resolution of imaging optics can be estimated by the Rayleigh criterion[1]

$$\sin\theta \propto \frac{\lambda}{A}, \tag{4.1}$$

where λ is the wavelength and A is the diameter of the aperture of the optics. In order to reduce the structure size on the microchips the wavelength has been reduced over the decades and microlithography is today performed at 193 nm, the wavelength of an ArF-excimer laser [10]. By using immersion liquids between the last lens and the wafer and by using double-patterning techniques today's microchips are produced as the 32 nm half-pitch node, which means that

---

[1]The same criterion, also known as the Abbe formula, had been published roughly 20 years earlier by Ernst Abbe [132].

4.1 Characterization of laser damage

the structures on the microchip have a spatial separation of 64 nm. These dimensions imply very high requirements for the imaging optics and therefore also for the lens materials. The materials need to have a very high homogeneity and the optical properties need to be stable under the application wavelength.

The alteration of the optical properties under irradiation is referred to as laser damage. In this chapter, I will discuss phenomenological observations of laser damage. If not stated otherwise, the experimental data in this chapter was provided by U. Natura.

4.1 Characterization of laser damage

Laser damage is characterized by a loss of the optical performance of the material, which is observed as a decrease of the transmission and the formation of scattering centers. For a high precision imaging process, scattering centers have a great effect on the imaging quality of the lenses. At the precision necessary for microlithography also absorption has a great influence as the thermal expansion of the material influences the refractive index.

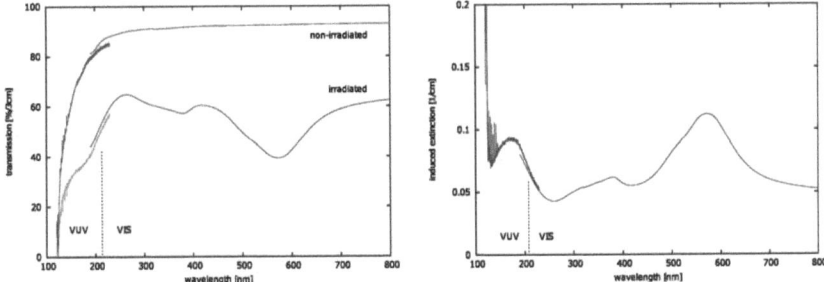

Figure 4.1: Induced extinction in CaF$_2$ [15]: Transmission of an irradiated and a non-irradiated area measured on the same sample (left). Induced extinction of the same sample (right).

An example for the change of a transmission curve after irradiation is shown in figure 4.1. In order to determine the effect of irradiation induced defects, it is necessary to observe the transmission before and after irradiation. The induced extinction is given by (3.14) and is also shown in figure 4.1. The induced extinction shows characteristic peaks which can be associated with different defect structures.

The radiation-induced defect structures in CaF$_2$ can be separated into two groups depending on the irradiation time necessary to create them in the material. The formation of point defects like F and H centers, but also M centers and Na stabilized F and M centers happen on a "short" timescale of about 20 minutes. This time may seem very long when compared to nearly all intrinsic timescales in solid state materials, but it is rather short compared to the expected lifetime of a high-performance optical lens and is therefore referred to as rapid damage. The formation of larger defect structures such as metallic colloids occurs on a much longer timescale. In long term measurements CaF_2 crystals are typically irradiated for about 2 months and this type of damage is referred to as long term damage.

4.2 Rapid damage

The rapid damage process in CaF_2 is often related to Na impurities, because when it was first investigated the stabilization of F centers by Na impurities was the dominant process for laser damage on the short timescale. Rapid damage related to Na impurities has been investigated thoroughly [36, 94, 14, 96, 17]. Mühlig et al. describe the formation of the optically active M_{Na} center from an F_{Na} center and a radiation induced F center and the radiation-induced annealing of these centers. They observed that the fluorescence of the M_{Na} center increases with rising temperature leading to the conclusion that the formation of M_{Na} centers is a diffusion driven process [17]. Because of the known Na-induced defect stabilization mechanism, great effort was made to push the impurity content of CaF_2 as low as possible (today it is below the ppb range), because it was expected that stabilized defect structures could thus be prevented.

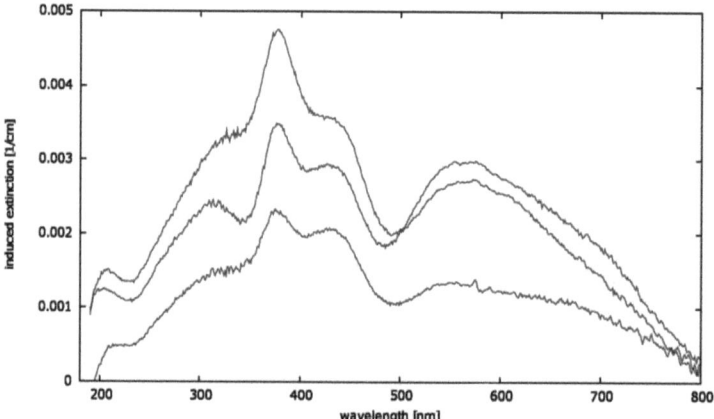

Figure 4.2: Typical induced extinction spectra of three CaF_2 samples after irradiation with 10^4 pulses at 50 mJ/cm². The peak at 378 nm is associated with F centers, the peak or shoulder at 310 nm with H centers. The peak at 200 and 580 nm could be allocated to beginning colloid growth (see chapter 3.5.3).

The extinction spectra of three typical CaF_2 samples are shown in figure 4.2. The samples were irradiated at 193 nm with 10^4 pulses and a fluence of 50 mJ/cm². The absorption of the

F center at 378 nm and of the H center at 310 nm are easily identified. The peaks at about 200 and 580 nm could be indicators for M center formation and beginning colloid growth. It can be concluded that after an irradiation time of about 20 minutes F and H centers have been formed in the material, and that first stabilization processes have taken place. After the transmission measurement the samples were exposed to UV light for about 15 minutes, and a complete annealing of the induced extinction was observed.

The annealing with radiation from a UV lamp is characteristic for rapid damage. Defect structures that cannot be annealed with radiation of a UV lamp are considered long-term damage effects.

4.3 Long term laser damage

Induced extinction measurements

For the characterization of long-term damage, four laser windows of different laser durabilities were irradiated in a LPX 220i Lambda Physik laser at 193 nm with up to $2 \cdot 10^8$ pulses at repetition rates of 60–100 Hz. The average energy density per pulse was 120 mJ/cm^2. The samples are labeled W1 through W4 where sample W1 has the highest laser durability and sample W4 the lowest. The results for the induced extinction are shown in figure 4.3.

The laser durability of a CaF_2 crystal is measured in a series of tests, which are not discussed here. They have been developed by former Schott Lithotec as a quality control tool, to ensure the performance of the produced crystals. The induced extinction after radiation decreases for the samples with higher laser durability, which was to be expected.

All samples exhibit the characteristic peaks just below 200 nm and around 550 nm allowing the conclusion that metallic Ca colloids have formed in the material (see chapter 3.5.3). The position of the peak below 200 nm cannot be determined exactly from these measurements, but the peak is clearly visible.

The position of the peak around 550 nm varies slightly for the different samples which can be attributed to different colloid sizes present in the material. For better comparison the data of all samples has been scaled to the maximum of the peak around 550 nm (figure 4.3). Due to the laser durability classification, a gradual change should be observable from sample W4 to W1. However, sample W3 shows unexpected behavior and I will therefore exclude it from my considerations for now.

With increasing laser durability the peak shifts from 550 nm to 575 nm. At the same time the extinction peak broadens significantly. In addition, the extinction at longer wavelengths increases relative to the peak. The shift of the peak position to longer wavelengths is an indicator that the colloids are larger. At the same time the broadening of the peak indicates that the size distribution of the colloids is broadened. The increased extinction at long wavelengths

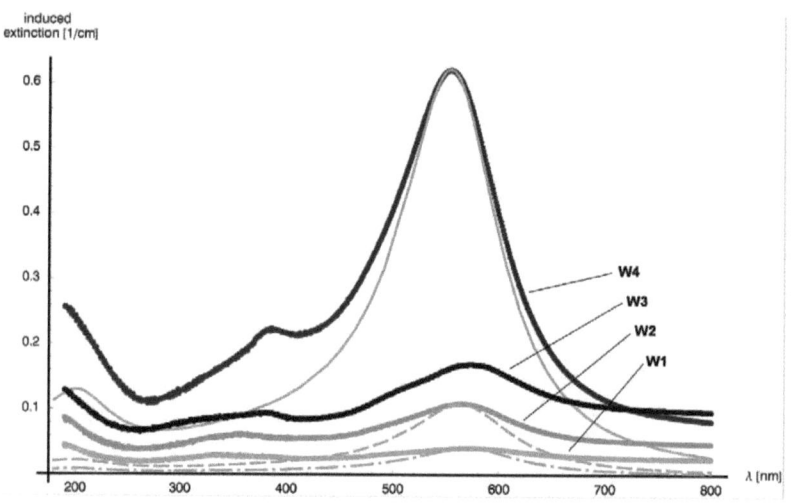

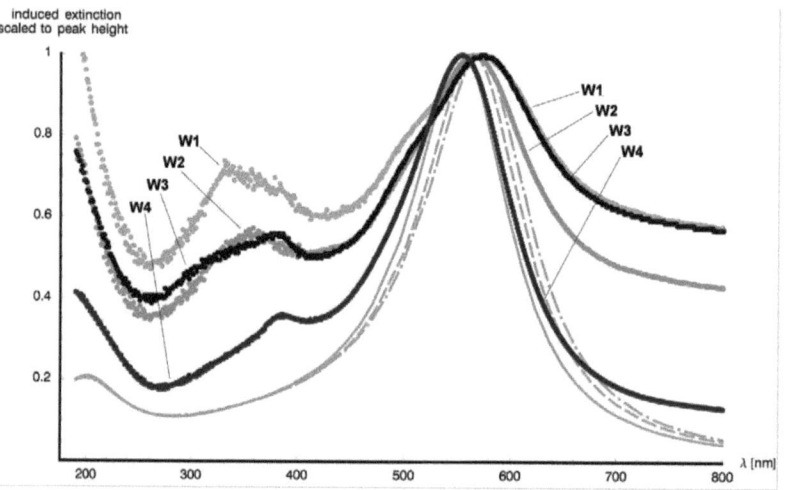

Figure 4.3: Top: Induced extinction of laser window samples W1–W4. The peaks below 200 nm and around 550 nm are characteristic for metallic calcium colloids in CaF_2. With increasing laser durability the induced extinction decreases. The theoretical extinction cross sections obtained from Mie theory were fitted to the data. For sample W4 the colloid radius r is 22 nm and the fractional colloid volume η is $2.8 \cdot 10^{-7}$. For sample W2 the fit parameters are $r = 26$ nm and $\eta = 5.1 \cdot 10^{-8}$ and for W1 $r = 28$ nm and $\eta = 2.0 \cdot 10^{-8}$. Sample W3 was not fitted because it showed slightly different behavior. Bottom: The same data and fits, but scaled to the maximum of the peak around 550 nm for better comparison.

relative to the peak could be an indicator for large colloids in the range of 50-100 nm.

For an estimate of the size of the colloids created in the irradiation process, the theoretical extinction obtained from Mie theory are fitted to the experimental data. For the fit the colloids were assumed to be monodisperse, i.e. they all have the same size. The two parameters used for fitting the induced extinctions are the colloid radius and the fractional colloid volume, i.e. the volume fraction of the material occupied by colloids. The colloid radius was used to fit the peak position and the fractional colloid volume was fitted to the height of the peak. The fits are shown in figure 4.3. For sample W4 the fitted colloid radius is 22 nm with a fractional colloid volume of $\eta = 2.8 \cdot 10^{-7}$. For samples W2 and W1, the fitted colloid radii are 26 nm and 28 nm, and the fractional colloid volumes are $\eta = 5.1 \cdot 10^{-8}$ and $\eta = 2.0 \cdot 10^{-8}$ respectively. Figure 4.3 shows that the fit for sample W4 is fairly good with respect to the shape of the peak. In order to account for the broadening of the peaks of samples W2 and W1, the peaks were fitted assuming a size distribution of the colloids, but a significant improvement of the fit was not observed. Nevertheless, based on the experiments I conclude that the size distribution of the colloids in sample W4 is very narrow and becomes wider for samples W2 and W1.

Sample W3 did not fit into the picture of a gradual change according to the laser durability classification. While the absolute induced extinction of sample W3 lies between that of sample W2 and W4 except for wavelength longer than 740 nm, the shape and position of the peak lies in between samples W1 and W2. At this point the reasons for this behavior are subject to speculation.

Further, samples W3 and W4 exhibit the F center peak at 378 nm, which is not visible in samples W1 and W2. Samples W1, W2, and W3 exhibit a slight shoulder at 520 nm, which is strongest in sample W1. In addition, samples W1 and W2 show a peak around 350 nm, which is slightly visible as a shoulder in sample W3 as well. The shoulder at 520 nm and the peak around 350 nm can be associated with the M center absorptions at 365 and 520 nm. Since the observed peaks are not exactly at 365 nm but a little below hat, I conclude that these M centers are either only loosely bound or associated with impurities. Rauch and Schwotzer [93] observed a shift of the absorption bands of the M center to smaller wave lengths for O^{2-} impurities. Even though impurities are an unlikely explanation due to the extremely high purity of the crystal, this could still be an explanation, as the presumable fractional colloid volume in sample W1 is in the range of 10 ppb and and at this level impurity concentrations in or below the ppb range could be visible in association with radiation induced point defects.

The M center peaks are visible best in the sample of highest laser durability and are not visible at all in sample W4. In sample W4 the colloid and F center peaks are much stronger than in the other samples and therefore most likely superpose the very small M center peaks.

The induced extinction measurements on irradiated laser windows have shown that metallic calcium colloids are formed in CaF_2 after irradiation with 193 nm. A higher quality of the

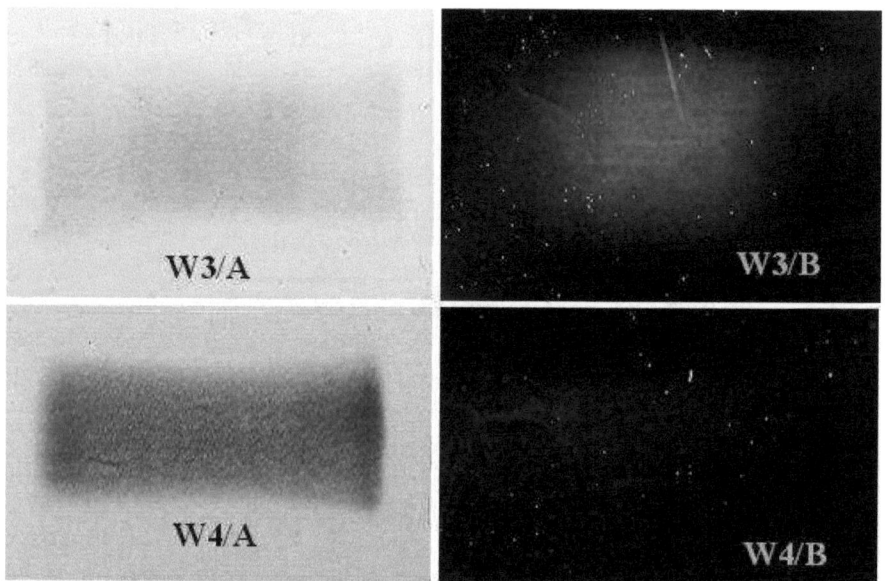

Figure 4.4: Absorption (A) and Scattering (B) of laser window samples W3 and W4. Sample W3 shows little absorption, but strong scattering while sample W4 shows strong absorption and little scattering.

material, which is in this case measured in terms of the laser durability, implies a lower absolute induced extinction. At the same time the size distribution of colloids is wider. Very good samples do not only exhibit the colloid peaks, but also M center related peaks.

Absorption and scattering

For samples W3 and W4, also absorption and scattering was observed (figure 4.4). While sample W3 shows hardly any absorption but strong scattering, sample W4 exhibits strong absorption but very little scattering. This observation first led me to the assumption that larger defect structures could be the cause for long term laser damage effects, because with a very similar extinction spectra, point defects should also show a similar absorption and scattering behavior. For larger defect structures on the other hand, the absorption and scattering behavior depends on the size of the structures.

In sample W4, the induced extinction measurements indicate smaller colloids with a narrower size distribution than in sample W3, and only little scattering is observed. However, it can be observed that the scattering colloids have a tendency to grow at lattice defects, such as small angle grain boundaries or line defects [133, 134]. Such lattice defects or in principle also impurities could be seeds for the growth of stable colloids, which is an indicator that colloids will start growing at defect sites with a larger probability.

Sample W3 shows strong scattering, from which I conclude that the colloids have grown larger than in sample W4. The extinction peak of sample W3 corresponds to colloids with a radius of 28 nm, which already show a strong increase in their scattering behavior compared to colloids with a radius of 22 nm (figure 3.22).

In addition, the purple coloration of sample W4 is typical for CaF_2 and is due to the F center absorption at 378 nm and the M center absorption at 365 nm, which is also observed in natural fluorite [23]. This is in good agreement with the induced extinction measurements (figure 4.3), which also shows the F center absorption at 378 nm in sample W4.

4.4 AFM measurements

Two CaF_2 samples were investigated with non-contact dynamic atomic force microscopy (NC-AFM) after irradiation. The measurements were performed by Prof. Michael Reichling at the University of Osnabrück [134].

Samples with a similar laser durability as samples W3 and W4 were chosen, because it was expected that a larger number of colloids could be observed. The samples were irradiated in the same way as the samples in long-term measurements. The irradiated samples were cleaved in an ultra high vacuum (UHV), where the AFM measurements were performed on the natural cleavage planes of CaF_2 ([111] plane). The measurements were repeated after two days showing no change of the observations. The results are shown in figure 4.5.

The AFM images show elevations and indents on the cleaved surface, which are below 10 nm in diameter and are distributed with a density of 10–20 per μm^2, giving a strong indication for colloid formation in CaF_2 [134]. In addition it was observed that no charges were present on the cleaved surface, as it is the case for non-irradiated CaF_2. This was taken as an indication that the charges were compensated by either the observed colloids or other defects such as F and M centers.

The results are a good confirmation of the assumption that Ca colloids are the defect structures formed after long-term irradiation.

4.5 Annealing by tempering

The conductivity measurements showed that F-H pairs are formed thermally above 200–230°C. As I will point out in section 5.6, the presence of equally distributed F and H centers should lead to an annealing of the colloids. If F-H pairs are formed statistically, this should therefore lead to an annealing of the colloids in the crystal above the formation temperature. Thus, I

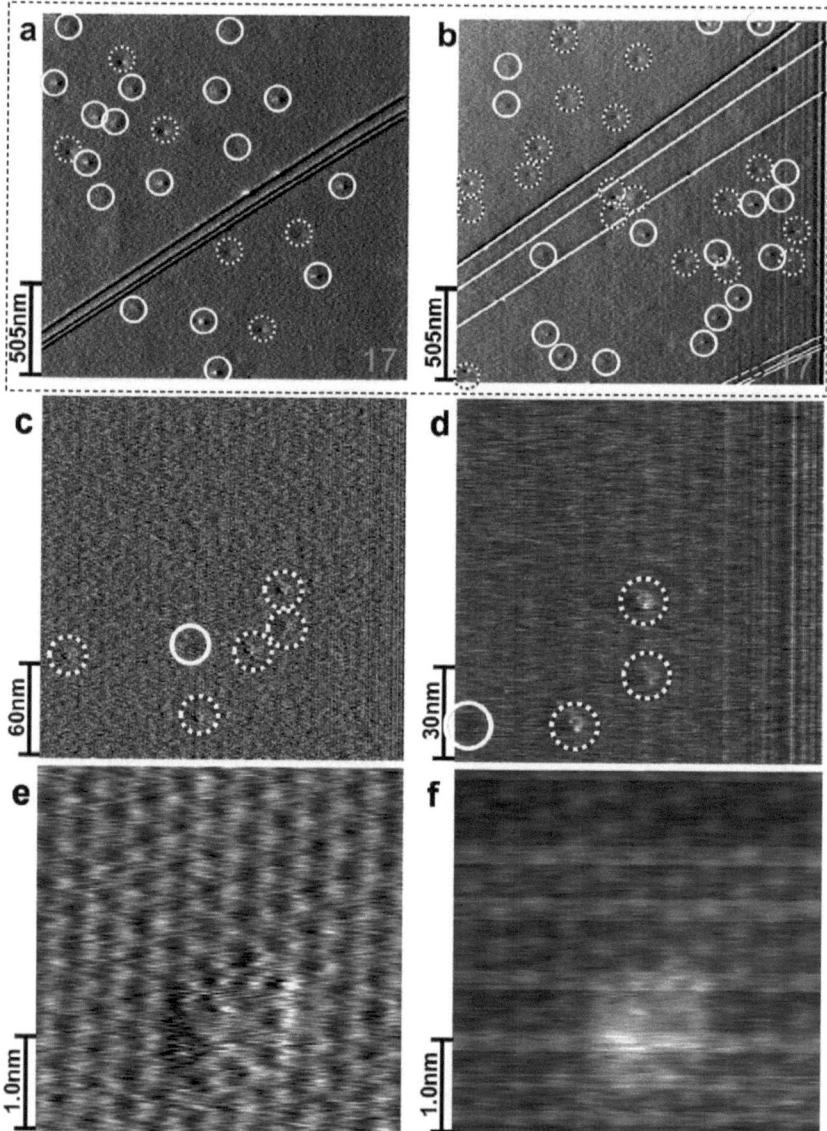

Figure 4.5: NC-AFM measurements performed by Reichling [134]. Images a and b show the distribution of colloids on the cleaved surface. In images c and d an rough upper limit of below 10 nm for the colloid size is observed. Images e and f show the cleaved surface with atomic resolution and subsurface colloids in the range of 1 nm.

4.5 Annealing by tempering

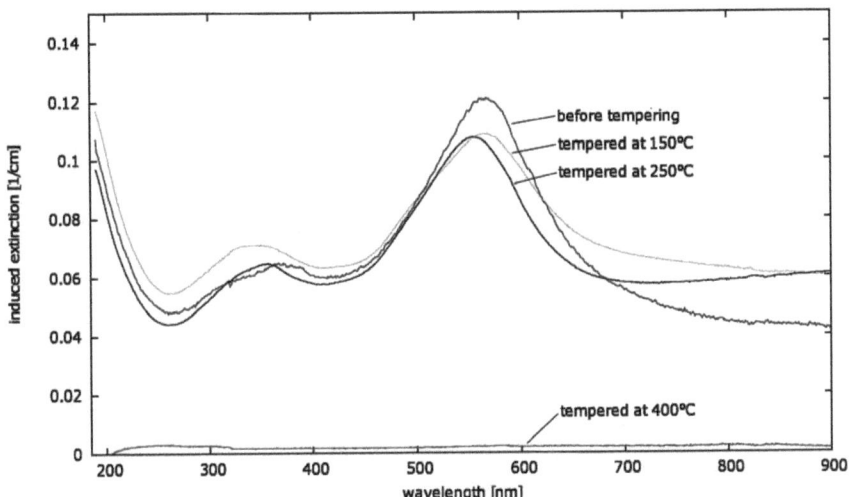

Figure 4.6: Annealing of CaF_2: The sample was tempered at 150, 250, and 400°C. The induced extinction profile shows slight changes after being tempered at 150 and 250°C, and vanishes after annealing at 400°C.

proposed annealing experiments to determine in what temperature range the annealing would take place.

A CaF_2 sample that showed a characteristic induced extinction spectrum was tempered at 150, 250 and 400°C for one week each. After each temper period the induced extinction was measured. The results are shown in figure 4.6.

After the first tempering period at 150°C, the colloid peak decreases and is broadened towards longer wavelengths. This is an indication that the colloids grow larger at these temperatures because of further defect agglomeration, which is due to the enhanced mobility of F centers at higher temperatures. Microscopically the colloids grow, as more F centers agglomerate on their surface. Additionally, the size distribution of the colloids could shift towards larger colloid sizes due to a process similar to Ostwald-ripening.

After tempering at 250°C the peak shifts slightly to the left and is narrowed, which can be interpreted as a decrease in size of the colloids. According to the conductivity measurements, a complete annealing of the colloids would be expected at this temperature because of the recombination with thermally created H centers. Since only a slight shrinking of the colloids is observed, I conclude that the rate at which F-H pairs are thermally formed is still too low at these temperatures to lead to a full annealing.

After tempering at 400°C, complete annealing of laser damage is observed in the material.

The growing and shrinking of the colloids during tempering at 150 and 250°C also suggests that the size of the calcium colloids depends on the thermal equilibrium of the defects in the material. Such a model has been proposed by Kuzovkov et al. [135].

Long term observation

A sample that showed characteristic laser damage after irradiation with $4 \cdot 10^8$ pulses at 50 mJ/cm^2 was stored for 8 months at room temperature. After this time period the induced extinction spectrum showed no sign of colloids any more, however the F center peak had grown with respect to the induced extinction measurements performed immediately after irradiation [136]. This can be seen as a hint that calcium colloids are not stable in CaF$_2$, but that their formation is enhanced by thermal and radiative activation.

4.6 Discussion of the observations

The formation of metallic Ca colloids in CaF$_2$ has been observed after irradiation with 193 nm laser radiation. The formation of Ca colloids in CaF$_2$ has been observed prior to this work [137, 138, 139, 140, 141, 142, 143], however in previous studies the colloids were induced by electron irradiation. The main difference between electron irradiation and laser irradiation is the locally deposited energy and the penetration depth of the radiation. The energy deposited locally by low energy electrons with 2 keV is much higher than the locally deposited energy of two 6.4 eV photons. In addition, the penetration depth of electrons is very small, for low energy electrons about 160 nm [140]. Irradiation with electrons deposits a large amount of energy in the surface-near region of samples, leading to much larger defect concentrations than after irradiation with photons.

In principle the dynamics of colloid formation should be similar after irradiation with electrons and after irradiation with photons. Since the defect concentration after electron irradiation is much higher, colloids will form faster. The size of the colloids will most likely depend on the amount of energy deposited which means that for a similar colloid formation the irradiation times for laser induced colloid formation are much longer, which is in good agreement with the timescales on which laser-induced formation of colloids is observed. The colloid size of 22 − 28 nm corresponds well with the observations by Bennewitz et al. [140, 141], who observed the formation of colloids after electron irradiation with mean colloid radii of 18 − 32 nm over a temperature range of 300 − 400K.

The annealing of colloids by tempering has been observed in CaF$_2$ irradiated with 0.5 − 1 MeV electrons [144] and in neutron-irradiated CaF$_2$ [145]. The conclusion that colloids are annealed at 250°C, which was deduced from the conductivity measurements in section 3.4.3, is well confirmed by Izerrouken et al. [145]. Unfortunately, they do not state the size of the

4.6 Discussion of the observations

colloids they observed. The experimental observations of this work are better confirmed by Wurster et al.[144], who observe annealing starting at 300°C after electron irradiation. It is a reasonable conclusion that the annealing process of Ca colloids in CaF_2 starts at around 250°C due to thermally induced F-H pairs, but the kinetics of the annealing of colloids are still very slow at these temperatures. This would be in agreement with the observations of annealing, where a decrease in the size of the colloids is observed after tempering at 250°C. I conclude that a complete annealing takes place at temperatures above 300°C in agreement with [144].

In addition to thermal annealing, Cramer et al. [142] reported the annealing of colloids after irradiation with 532 nm laser radiation, which is interpreted as a local heating of the colloids by the absorbed radiation, which leads to the "thermal" decomposition and annealing of the colloids. This observation could also explain the finding of only small colloids in the AFM measurements. Since the colloids are formed during radiation and exhibit an absorption peak around the application wavelength of 193 nm, I conclude that small colloids are formed during the irradiation process, but that their growth is limited by the irradiation which they absorb at the same time. This also implies that the growth of larger colloids occurs after irradiation in a process similar to Ostwald-ripening or by high local energy deposition.

In electron-irradiated CaF_2 the desorption of gaseous fluorine from the sample surface has been observed [146]. In the irradiation experiments fluorine desorption from the crystals was not observed. The main reason is that the damaging process is observed in the bulk of the material and not in a surface near region. Since the colloids are an agglomeration of F centers and the F centers are formed in pairs with the H centers, the interstitial fluorine ions need to be absorbed in the material as well. If the defects are created in a surface near region, desorption of fluorine could be possible. However with the focus on defect formation in the bulk, desorption of fluorine from the crystals is not expected, but the created interstitials are expected to remain in the material in the form of H centers.

Chapter 5

A laser damage model

5.1 The pre-irradiation material

Prior to any radiation I assume a perfect CaF_2 crystal. However, a real material at finite temperatures shows intrinsic defect concentrations of F and H centers with an excess of F centers. At 300 K these defect concentrations have been determined to be $9.74 \cdot 10^{-19}$ ppm for the F centers and $6.84 \cdot 10^{-119}$ ppm for the H centers [118]. However at higher temperatures these defect concentrations increase significantly. At 1500 K the defect concentrations are 5.04 ppm and $4.70 \cdot 10^{-19}$ ppm for the F and H center respectively, at 1650 K they are 13.4 ppm and $6.72 \cdot 10^{-17}$ ppm [118]. Since synthetic crystals are grown from the melt and later tempered in a fluorine atmosphere, it can be assumed that the defect concentrations equilibrate at the values at room temperature. These defect concentrations are so small that I will neglect them for reasons of simplicity.

5.2 Irradiation induced defect formation

CaF_2 has a band gap of 11.8 eV, and is therefore transparent at 193 nm, which corresponds to 6.4 eV. Assuming an ideal CaF_2 crystal, free of any defects or impurities, electromagnetic radiation is absorbed via a two-photon process [147, 148]. The probability for this process is small but finite. Since the probability for two-photon absorption increases quadratically with the irradiation intensity, two-photon absorption processes become important at high laser irradiation densities, as they occur in microlithography. In principle multi-photon absorption processes can also contribute, but their probability is negligible since it scales with the nth power of the intensity.

The two-photon absorption process creates a free electron-hole pair. The electron-hole pair then relaxes via different lattice distortions [103, 106, 102, 107]. First, the hole localizes as a V_k center, which is an F_2^- complex oriented along the (100) direction. In a second step the

5.2 Irradiation induced defect formation

electron localizes on one of the fluorine site of the V_k center forcing the F_2^- complex to rotate to the (111) direction. The result is the so-called self-trapped exciton (STE), which can also be considered as a nearest-neighbor F-H pair in halide crystals with the fluorite structure. The STE in CaF_2 can recombine via an optically forbidden transition emitting at 278 nm, which has a fairly large lifetime of 1.1 μs at room-temperature [107, 94, 109]. Instead of recombination there is also the chance of the separation of the F center and the H center leading to a stable F-H pair. Only about 1% of the STEs evolve into stable F-H pairs [102, 106]. However, these stable F-H pairs are the lattice deformations that remain in the crystal as laser induced defects.

In its application in the semiconductor industry, CaF_2 is exposed to extremely high radiation densities. Since absorption is only possible via the two-photon process mentioned above, the formation rate of F-H pairs will be proportional to I^2, where I is the radiation intensity. Due to the relatively large lifetime of the F-H pair, the concentration of F and H centers in the material will increase gradually. The recombination rate of F and H centers will scale with the concentrations of the defects c_F and c_H. After a certain time these defect concentrations will reach a saturation level, which depends on the incident radiation density. This has been concluded from experimental results [94].

It is important to note that the 193 nm ArF excimer laser is a pulsed laser and that the pulse length and shape as well as the laser frequency will have an effect on the defect formation and recombination rates. Because the initial defect formation is proportional to I^2, a flat stretched pulse will create less defects than a sharp short pulse. In addition, longer relaxation times between pulses will allow for more recombination events between two pulses leading to a slower defect formation in the long run. The effect of the pulse shape and length will not be considered explicitly in this work, but is topic of on-going research [149].

In principle, no laser damage would occur if all the F and H centers recombined, as this would lead back to the undamaged state of the crystal. However, at high radiation densities a large number of F-H pairs is created, increasing the defect concentration in the irradiated bulk. This induces a concentration gradient at the boundary between the irradiated and the non-irradiated bulk, resulting in migration of point defects out of the irradiated bulk into the non-irradiated bulk. Since the diffusion barrier of the H centers is much lower than the barrier of the F centers, this leads to an F center excess in the irradiated bulk, while the H centers are distributed throughout the non-irradiated crystal. This idea is schematically illustrated in figure 5.1.

These assumptions can be described by rate equations for F and H center concentrations c_F and c_H

$$\frac{\partial c_F}{\partial t} = A_0 I^2 - A_R c_F c_H - \nabla \cdot \mathbf{J}_F, \tag{5.1}$$

$$\frac{\partial c_H}{\partial t} = A_0 I^2 - A_R c_F c_H - \nabla \cdot \mathbf{J}_H, \tag{5.2}$$

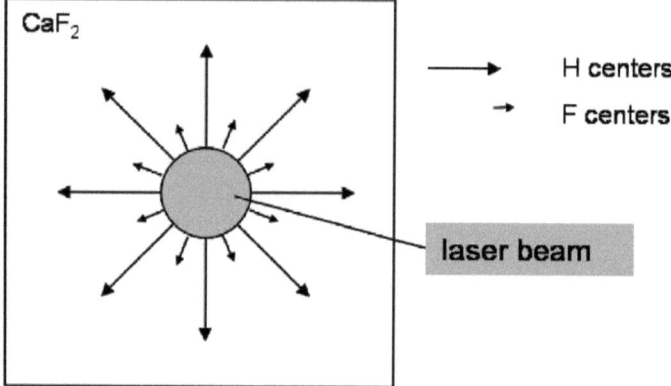

Figure 5.1: Depicted diffusion model of laser damage. The slice shown is perpendicular to the direction of the laser beam. The radiation induced concentration gradient leads to a diffusion of point defects to non-irradiated ares of the crystal. Since the diffusion barrier of the H centers is lower than that of the F centers, the F center concentration in the irradiated area will be higher than the H center concentration.

where A_0 and A_R are the coefficients for the F and H center formation and recombination and J_F and J_H are the fluxes of F and H centers respectively. These equations describe the defect concentrations in the irradiated bulk. Up to this point any defect stabilization has been neglected.

5.3 Defect agglomeration

The radiation-induced point defects can be stabilized by impurities or by agglomeration. Due to the high purity level of CaF_2 I will focus on defect-defect stabilization and neglect any impurity effects. My calculations have shown that an M center is energetically favored in comparison to two F centers (see table 3.2). This implies that two F centers can stabilize themselves by forming an M center. The experimental observations also show the formation of metallic calcium colloids, which means that F centers cannot only stabilize by M center formation but also by further agglomeration.

From an energetic point of view the recombination of F and H centers is ideal. However, due to the diffusion of the defects the concentration of F centers in the irradiated area is higher than the H center concentration. Therefore, not all F centers can recombine with H centers and they start agglomerating in the irradiated crystal bulk.

A new term is introduced to describe the stabilization and agglomeration of F centers. Since the formation of M centers is proportional to c_F^2 and the formation rate of colloids will also depend on the F center concentration, I denote the stabilization of point defects by agglomeration as $A_{agg}c_F$ and my rate equations yield

$$\frac{\partial c_F}{\partial t} = A_0 I^2 - A_R c_F c_H - \nabla \cdot \mathbf{J}_F - A_{agg}(c_F, \eta)c_F, \tag{5.3}$$

$$\frac{\partial c_H}{\partial t} = A_0 I^2 - A_R c_F c_H - \nabla \cdot \mathbf{J}_H - A_{ann}(\eta)c_H, \tag{5.4}$$

where a term for the annealing of the stabilized defect structures A_{ann} has also been added. The coefficients of the agglomeration and the annealing rate will depend on the concentration of stabilized defect structures, which has been introduced as the fractional colloid volume η, which is the volume fraction occupied by colloidal structures. I regard all stabilized defect structures in the material as immobile, and have included the term for M center formation in the agglomeration term because M center formation is the initial step towards colloid formation. Thus the M center concentration is also included in the fractional colloid volume η.

5.4 Diffusion based laser damage model

For clearness I will distinguish two steps in the formation process of stable defect structures in the crystal. The first step is the rapid damage process. At the beginning I assume that the crystal is perfect. Under irradiation F-H pairs are formed until the formation rate and the recombination rate are approximately the same. For simplicity, I will assume that agglomeration of F centers does not occur during this process. In a real material F center stabilization can also occur already at this early stage. However, as long as the local concentration of F and H centers is the same, recombination of F and H centers will be the preferred stabilization mechanism, because the perfect crystal is always preferred to a defect structure.

After irradiation with 30,000 pulses an equilibrium concentration of F and H centers has been induced in the material [94]. Under the assumption that no F center stabilization takes place yet, that the formation and recombination rates are in equilibrium, and that F and H centers do not interact, the rate equations reduce to

$$\frac{\partial c_F}{\partial t} = -\nabla \cdot \mathbf{J}_F, \tag{5.5}$$

$$\frac{\partial c_H}{\partial t} = -\nabla \cdot \mathbf{J}_H, \tag{5.6}$$

which according to Fick's second law can be written as

$$\frac{\partial c_F(\mathbf{r},t)}{\partial t} = -D_F \Delta c_F(\mathbf{r},t), \tag{5.7}$$

$$\frac{\partial c_H(\mathbf{r},t)}{\partial t} = -D_H \Delta c_H(\mathbf{r},t), \tag{5.8}$$

where Δ is the Laplace operator. The simplifications lead to independent diffusion equations for the F and H center. For the solution of these diffusion equations, I assume that the defect concentration c_0 in the irradiated area is constant, because new defects are continually produced and their concentration will be in an equilibrium due to recombination, diffusion and stabilization. Additionally, I assume that the initial concentration outside the irradiated area c_{ini} is zero for F and H centers and that the diffusion is not limited by the size of the crystal, i.e. at infinite distance the concentration is zero. I assume cylindrical symmetry with the z-axis pointing along the direction of the laser beam. With these boundary conditions the solution of the diffusion equations results to

$$c_i(r,t) = c_0 + (c_0 - c_{\text{ini}})\frac{2}{\pi}\int_0^\infty e^{-D_i u^2 t}\frac{J_0(ur)Y_0(ar) - Y_0(ur)J_0(ar)}{J_0^2(ua) + Y_0^2(ua)}\frac{du}{u} \qquad (5.9)$$

for $r > a$, where the index i indicates the F or H center, a is the radius of the laser beam, i.e. the irradiated area, D_i is the respective diffusion coefficient, and J_0 and Y_0 are the 0th order cylindrical Bessel functions of the first and second kind respectively.

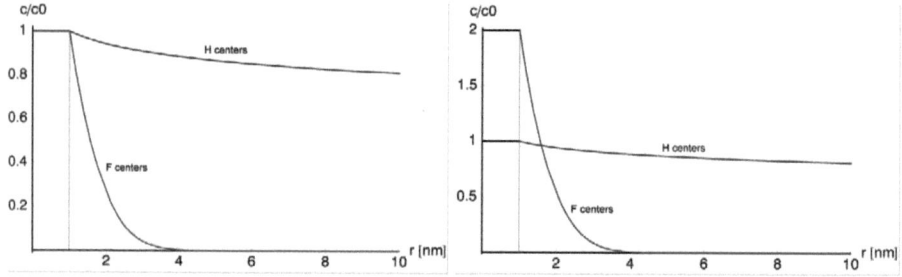

Figure 5.2: Schematic representation of the concentration profile as a function or the radius for a timescale of 1 month for F (red) and H (blue) centers at 300 K. A constant concentration was assumed in the area of irradiation ($0 < r < 1$). Left: After the rapid damage process an equal amount of F and H centers are present in the material. The H centers diffuse much faster than the F centers. This leads to a quick increase of the F center concentration in the irradiated area of the material, which is shown on the right. Right: In principle the spacial integrals over the concentrations should be equal at all times. This effect is schematically depicted with an increased F center concentration in the irradiated area.

The solution for F and H centers for two different cases on the timescale of one month is shown in figure 5.2. Since the diffusion process takes place in an ionic crystal with high energy barriers, it is not surprising that the timescale is so large compared to the length scale in the range of nm. The left image shows the concentration profiles after the rapid damage process, when the concentrations of the F and H centers in the irradiated area are equal. With a constant concentration of F centers in the irradiated area the diffusion length after one month is below 4 nm for the F centers, while the H centers diffuse into the non-irradiated crystal

5.4 Diffusion based laser damage model

very quickly. Therefore, the concentration of F centers in the irradiated area will be much higher than the concentration of H centers. This concentration difference is shown in the right image. Under continuous irradiation the F center concentration in the irradiated area will increase more and more, while the H centers diffuse into the non-irradiated area. Under these conditions F centers will agglomerate and form stable defect structures such as M centers, but under continuous irradiation also larger agglomerates up to Ca colloids.

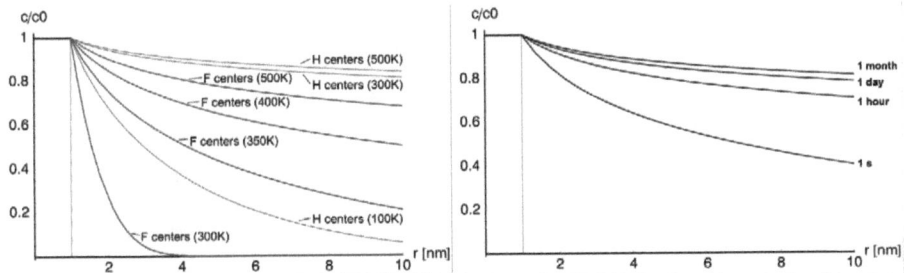

Figure 5.3: Left: Concentration profiles of F (red) and H (blue) centers for different temperatures after a diffusion time of 1 month. Left: Concentration profiles of the H center for different diffusion times at 300 K.

The concentration profile of the H centers after 1 s, 1 hour, 1 day, and 1 month are shown in right image in figure 5.3. Already after 1 s a large amount of H centers has migrated several nm into the non-irradiated material. This implies that the concentration difference in the irradiated area due to defect diffusion will already occur during the rapid damage process. Therefore, I conclude that even the formation of M centers only takes place due to a difference of the F and H center concentrations.

In addition, the temperature dependence of the diffusion process is shown in the left image. For increasing temperatures up to 500 K the H center concentration does not show significant change. The F center diffusion on the other hand increases strongly coming close to the concentration of the H center. While the F center does not show significant diffusion at 100K, the diffusion of the H centers at this temperature is considerably lower.

First, I conclude that the mobilities of both F and H centers play a significant role in the annealing process of laser damage because an increased mobility allows for a better recombination. Second, I conclude that for temperatures around or below 100 K the stabilization of defects is not as strong because the difference of the concentrations of F and H centers will not be as large as at room temperature due to the slower diffusion of the H centers in this temperature range. Last, the difference of the concentrations of F and H centers is also not large for higher temperatures, but here the reason is that the diffusion of the F centers is in a similar range as the diffusion of the H centers leading to similar concentrations throughout the crystal and thus allowing for recombination.

5.5 Limits of stability of Ca colloids in CaF_2

In section 3.6 I have shown that colloids will grow as long as there are F centers present. However, after irradiation colloids with a finite size of up to 30 nm are observed. This size was deduced from the induced extinction measurements and in agreement with other discussed results. In AFM measurements the colloids were well below 10 nm in size.

Because Ca colloids have an absorption peak in the range of 193 nm, growing colloids absorb radiation energy during the irradiation process. The formation energy of a colloid with a radius of 1 nm is 75 eV according to my calculations (see chapter 3.6.4). During the irradiation with a laser pulse with a fluence of 120 mJ/cm² this colloid absorbs $3.6 \cdot 10^3$ eV (calculated from equation 3.22). The absorption of radiation will lead to a heating of the colloids and the absorbed radiation supplies much more energy than the formation energy of the colloids, which then leads to the dissolving of the colloids. Thus, I conclude that the formation of colloids takes place after completion of the radiation process or in between pulses. Since smaller F center agglomerates such as the M center do not absorb at 193 nm, these stabilized defect structures will be formed even during the irradiation process.

For the experimental setup, where the crystals are irradiated continuously, I conclude that the formation of colloids will take place after the completion of the irradiation time. However, after the completion of the irradiation process the formation of colloids will take place within a short period of time because of the very high F center concentration in the irradiated area. The growth of the colloids is only limited by the F center concentration. During tempering of the material below 250°C the colloids can grow similar to Ostwald-ripening resulting in larger but fewer colloids.

5.6 Reversibility of laser damage

After long-term irradiation, stabilized F centers and colloids in the irradiated crystal and H centers in the surrounding non-irradiated crystal are present. In order to reverse the laser damage process, the defects need to recombine, so that the defects are annealed and the perfect crystal lattice is restored. This can be done by thermal annealing above $250 - 300$°C recovering the original optical properties of the material. I propose the following mechanism for the reversing of laser damage in a tempering process.

Above a threshold temperature F-H pairs are thermally induced everywhere in the crystal, leading to a uniform distribution of F and H centers on top of the radiation induced defect concentrations within the crystal. For higher temperatures the number of thermally induced

F-H pairs obviously increases. In addition to the thermally induced uniform F and H center concentrations, the mobility of the F centers is much higher at this temperature, which leads to a recombination of F and H centers, that is not limited to their formation sites because both defects are mobile at the annealing temperature. After a certain annealing time all defects have recombined, and the defect-free crystal lattice is restored.

5.7 Preventing laser damage

From the viewpoint of the application the primary question is not how the laser damage process works, but rather how it can be prevented or what can be done against it. I present three ideas to prevent the damaging of CaF_2 lenses by 193 nm laser irradiation.

The general idea of the prevention of laser damage is to maintain a locally equal concentration of F and H centers in the material, so that they can recombine rather than form large stable defect structures in the material.

The first approach is to cool the lenses during irradiation, so that the H centers become immobile and do not migrate to the non-irradiated crystal areas. Since the H centers become mobile at 170K [34], this requires the cooling of the complete optics below this temperature. Because the lenses for microlithography have a diameter of up to 35 cm this is probably not a feasible option.

The second approach is to heat the lenses during irradiation, so that the F and H centers are mobile. However, the H centers will migrate faster than the F centers. Therefore, this approach will strongly reduce laser damage, but not fully prevent it. Above the annealing temperature this approach should eliminate any laser damage. Nevertheless, the heating of the whole optical system is not practical for the same reason as the cooling.

My third proposition is to irradiate the entire crystal from time to time to uniformly induce F-H pairs, which allow for the full recombination of defects. The mechanism is similar to thermal annealing except that the F-H pairs are induced by irradiation. At room temperature the F centers are not mobile, but if the entire crystal is irradiated every few seconds this should not influence the results. This idea is also practicable in the sense that it could be implemented in a wafer stepper. The irradiative annealing could take place after each step or after each wafer is finished, depending on how strong the rapid damage effects in a real material are in the first few seconds.

Chapter 6

Conclusion

High purity single-crystalline CaF_2 is one of the key lens materials in microlithography. The excellent performance of CaF_2 is mainly due to its high laser durability. It is of crucial importance that the optical properties of the lens material are maintained during the manufacturing process and are not altered by the radiation of the application itself. In the past decade great efforts have been made to produce CaF_2 of very high purity and laser damage of the material could be reduced. However, even in CaF_2 of highest purity with an impurity content below the ppb range laser damage effects are still observed and an understanding of laser damage processes in CaF_2 is necessary.

It is well known that the irradiation with 193 nm induces F-H pairs in CaF_2 and that the stabilization of F centers in the form of M centers plays an important role in understanding laser damage in CaF_2, which is often related to Na^+ impurities. However, CaF_2 samples of very high purity, which did not show any sign of Na^+-related defect stabilization, still exhibited laser damage effects, meaning that an intrinsic defect stabilization process must take place in CaF_2.

The intrinsic stabilization mechanism was identified as the formation of metallic Ca colloids. Because F centers self-stabilize as M centers the further agglomeration of F centers was a obvious hypothesis, which was found to be in excellent agreement with experimental results. The formation of Ca colloids in CaF_2 explains the intrinsic defect stabilization in CaF_2 leading to the necessary understanding for the underlying defect structures.

The approach of this work is to provide a theoretical description of the relevant properties of point defects and the other defect structures related to laser damage in CaF_2. I used electronic structure calculations to determine the properties of the relevant point defects in CaF_2, the F center, H center and M center. Because of the former importance of Na^+ impurities, I also looked into the stabilization of the point defects by Na^+ impurities. In addition to the monovalent Na^+, I also investigated the effect of the trivalent Y^{3+} as a cationic impurity.

The properties of the M center formation in pure CaF_2 compared to the formation in CaF_2

with Na$^+$ impurities suggested that the stabilization of point defects was a dynamical, diffusion driven process, which was also confirmed experimentally. Even though the diffusion properties of the F and H centers had been investigated, the values are widely scattered. I calculated the diffusion coefficients for the F and H center using *ab initio* quantum mechanical methods with the nudged elastic band method to determine the minimum energy path for the diffusion processes and to determine the diffusion barrier. The phonon modes of the F and H center were calculated to determine the prefactors of the diffusion coefficients. The diffusion properties and the mobility of the F and H centers play an important role in the explanation of laser damage.

The optical properties of metallic Ca colloids in CaF$_2$ were calculated from Mie-theory showing good agreement with experimental results, which allowed me to conclude that Ca colloids have been formed in the material. The formation of metallic colloids in a crystalline matrix is usually rather unlikely, but the unique property that the structure of the Ca atoms in metallic Ca and CaF$_2$ is the same and that the lattice constants are almost identical, provides the setting for such an unusual effect.

With the purpose of determining an upper limit for the stability of colloids in CaF$_2$, I calculated the formation energy of the colloids. My original assumption that the pressure exerted on the colloids by the surrounding matrix would limit the growth of the colloids did not hold true. Thus, the growth of Ca colloids is in principle only limited by the availability of F centers. However, since the colloids absorb at 193 nm their growth is also limited by the absorbed energy during the irradiation process, which does not allow the growth of large colloids during the irradiation process.

On the basis of my results for the properties of the point defects and Ca colloids, I propose a diffusion-based laser damage model for CaF$_2$. At room temperature F-H pairs are induced by 193 nm laser irradiation. A part of the induced H centers migrates to non-irradiated areas of the crystal in a concentration gradient driven diffusion process, which leads to a large number of F centers in the irradiated area which cannot recombine with H centers and therefore start to self-stabilize and agglomerate forming M centers and eventually Ca colloids. Because the growth of colloids is limited during the irradiation process due to the absorption of the colloids at 193 nm, colloid formation most likely takes place after the irradiation process. Because of the high F center concentration, relatively fast colloid formation is expected after irradiation with a large number of pulses.

My model includes a mechanism for the annealing of colloids by tempering. Tempering above 250 − 300°C will thermally induce F-H pair formation, inducing a uniform F and H center concentration in addition to the radiation induced concentration profiles. Since both defects are mobile at these temperatures, the total F and H center concentrations throughout the crystal will equilibrate, so that all F and H centers can recombine in the annealing process.

Based on the understanding of the laser damage mechanism on a microscopic level, I propose three measures for the prevention of laser damage of which I regard two as not feasible. The basic idea is to maintain a locally equal concentration of F and H centers during the irradiation process so that all F and H centers can recombine after irradiation. This can be achieved by either limiting the mobility of the H centers, i.e. cooling, increasing the mobility of the F centers, i.e. heating, or by creating F and H centers uniformly in the crystal by radiation in an appropriate time interval. The measures for the prevention of laser damage are so far suggestions, which result from my model for laser damage in CaF_2. The prevention measures have not been tested experimentally, but I am confident that this can be done successfully.

In addition to the explanation of laser damage in CaF_2, this work can also be seen as a principle study of radiation damage in optical materials. The performance of any optical material strongly depends on the stability of its properties especially in its application environment. From a theoretical point of view CaF_2 is a very simple material because it contains only two element species, which are ionically bound and the fluorite structure is a simple structure with high symmetry. This makes a theoretical approach much easier than in more complex crystalline structures or even amorphous structures such as glasses, where the damage process is referred to as solarization. In addition, the high purity of CaF_2 today allows to access the intrinsic stabilization mechanisms experimentally. An understanding of the underlying processes of radiation damage is crucial for the improvement of optical components for future applications.

While my laser damage model has been developed specifically for CaF_2, it has many aspects which can be generalized to radiation damage processes in any optical material. Every highly transparent optical materials absorb a small fraction of the radiation. The absorbed photons cause an electronic excitation in the material, which can remain stable in the material. In a crystalline material this process occurs via the formation of an electron-hole pair, which then stabilizes in the material, such as the self-trapped exciton in CaF_2. In amorphous systems such as fused silica (amorphous SiO_2), this process occurs as the breaking of strongly polarized bonds [150]. This charge separation or charge transfer is the first step towards radiation damage in a material and is important for the understanding of radiation damage processes.

The stabilization of radiation induced defects is the second step in the radiation damage process. In CaF_2 this occurs as stabilization by impurities, self-stabilization and colloid formation. This process is strongly influenced by the mobility of the radiation induced defects. The stabilization process of radiation induced defects is the crucial issue in understanding the damage mechanism. CaF_2 is an ideal model system because the stabilization mechanism is simple. In more complex structures the stabilization mechanisms will be very difficult to identify. Today's optical materials often contain a multitude of components, whose combination makes the identification of defect stabilization almost impossible. Therefore, the understanding of radiation damage in glasses, which is referred to as solarization, can only be accessible through an approach via the simplest amorphous optical materials. Once the simplest case has been understood, the influence of additional components can be investigated.

Concluding, my model is able to explain laser damage in CaF_2 and it has the potential to be generalized to optical damaging processes in general. The laser damage process as such is a very complex dynamical process of defect formation, recombination, stabilization and diffusion in the material. I have described the aspects of the laser damage process with some simplifications, and created a consistent model for laser damage and the annealing of damage, resulting in a suggestion to prevent laser damage.

Appendix A

Cross sections from Mie-theory

In 1908 Gustav Mie developed the basic equations for what is today called Mie-theory [21]. At the time his work was motivated by understanding coloration effects in minerals and colloidal solutions. Mie-theory yields the absorption, scattering, and extinction cross sections of spherical particles with a complex refractive index surrounded by a transparent matrix. These results are obtained from the Maxwell equations where the only input is the refractive indices of the two media and the size of the spherical particle.

The basic equations

I will roughly sketch the solution yielding the basic equations of Mie-theory. I will follow the derivation in [151] only summarizing the main results.

The basic procedure is to solve the Maxwell equations for an incident plane wave in a medium with a real refractive index n_m and a spherical particle with a complex refractive index n_s, yielding the extinction and scattering cross section of the particle.

Solution of the vector wave equation in spherical coordinates

In a linear, isotropic, homogeneous medium the electromagnetic fields must satisfy the Maxwell equations

$$\begin{aligned} \nabla \cdot \mathbf{E} &= 0, \\ \nabla \cdot \mathbf{H} &= 0, \\ \nabla \times \mathbf{E} &= i\omega\mu\mathbf{H}, \\ \nabla \times \mathbf{H} &= -i\omega\epsilon\mathbf{E}, \end{aligned} \quad (A.1)$$

$$(A.2)$$

which implies the vector wave equations

$$\nabla^2 \mathbf{E} + k^2 \mathbf{E} = 0, \qquad \nabla^2 \mathbf{H} + k^2 \mathbf{H} = 0, \qquad (A.3)$$

where $k^2 = \omega^2 \epsilon \mu$. Equations (A.1) are fulfilled by constructing a vector function $\mathbf{M}$

$$\mathbf{M} = \nabla \times (\mathbf{c}\psi), \qquad (A.4)$$

where ψ is a scalar function and $\mathbf{c}$ is an arbitrary constant vector. Evaluating equation (A.3) for $\mathbf{M}$, shows that $\mathbf{M}$ satisfies the vector wave equation if ψ is a solution to the scalar wave equation

$$\nabla^2 \psi + k^2 \psi = 0. \qquad (A.5)$$

A second vector function $\mathbf{N}$ is now constructed

$$\mathbf{N} = \frac{\nabla \times \mathbf{M}}{k}, \qquad (A.6)$$

which also satisfies the vector wave equation. For symmetry reasons $\mathbf{c} = \mathbf{r}$ is chosen, because this leads to $\mathbf{M}$ being perpendicular to any sphere with $|\mathbf{r}| = $ constant.

The scalar vector wave equation in spherical coordinates is

$$\frac{1}{r^2}\frac{\partial}{\partial r}\left(r^2 \frac{\partial \psi}{\partial r}\right) + \frac{1}{r^2 \sin\theta}\frac{\partial}{\partial \theta}\left(\sin\theta \frac{\partial \psi}{\partial \theta}\right) + \frac{1}{r^2 \sin\theta}\frac{\partial^2 \psi}{\partial \phi^2} + k^2 \psi = 0. \qquad (A.7)$$

Solutions of the form

$$\psi(r,\theta,\phi) = R(r)\Theta(\theta)\Phi(\phi), \qquad (A.8)$$

lead to three separated equations

$$\frac{d^2 \Phi}{d\phi^2} + m^2 \Phi = 0, \qquad (A.9)$$

$$\frac{1}{\sin\theta}\left(\sin\theta \frac{d\Theta}{d\theta}\right) + \left[n(n+1) - \frac{m^2}{\sin\theta}\right]\Theta = 0, \qquad (A.10)$$

$$\frac{d}{dr}\left(r^2 \frac{dR}{dr}\right) + \left[k^2 r^2 - n(n+1)\right]R = 0, \qquad (A.11)$$

where the separation constants m and n are determined by subsidiary conditions that ψ has to satisfy. The solutions ϕ_m and ϕ_{-m} are not linearly independent. The independent solutions are

$$\phi_e = \cos m\phi \quad \text{and} \quad \phi_o = \sin m\phi \qquad (A.12)$$

for even and odd values of m.

The solution of the scalar wave function is given by

$$\psi_{emn} = \cos m\phi \, P_n^m(\cos\theta) \, z_n(kr), \tag{A.13}$$

$$\psi_{omn} = \sin m\phi \, P_n^m(\cos\theta) \, z_n(kr), \tag{A.14}$$

where $P_n^m(\cos\theta)$ are the associated Legendre functions of the first kind of degree n and order m and z_n is any of the four spherical Bessel functions j_n, y_n, $h_n^{(1)}$, or $h_n^{(2)}$. The solutions for $\mathbf{M}$ and $\mathbf{N}$ are then given by

$$\mathbf{M}_{emn} = \nabla \times (\mathbf{r}\psi_{emn}), \quad \mathbf{M}_{omn} = \nabla \times (\mathbf{r}\psi_{omn}), \tag{A.15}$$

$$\mathbf{N}_{emn} = \frac{\nabla \times \mathbf{M}_{emn}}{k}, \quad \mathbf{N}_{omn} = \frac{\nabla \times \mathbf{M}_{omn}}{k}. \tag{A.16}$$

Expansion of a plane wave in vector spherical harmonics

The fields inside the sphere E_1 and H_1 and the fields in the surrounding medium E_2 and H_2 need to be considered, where the fields in the surrounding medium can be written as the sum of the incident and the scattered field

$$E_2 = E_i + E_s, \quad H_2 = H_i + E_s. \tag{A.17}$$

The incident plane wave is arbitrarily chosen to be x-polarized. The incident field E_i in spherical coordinates is given by

$$\mathbf{E}_i = E_0 \, e^{ikr\cos\theta} \, \hat{\mathbf{e}}_x, \tag{A.18}$$

where

$$\hat{\mathbf{e}}_x = \sin\theta\cos\phi\,\hat{\mathbf{e}}_r + \cos\theta\cos\phi\,\hat{\mathbf{e}}_\theta - \sin\phi\,\hat{\mathbf{e}}_\phi. \tag{A.19}$$

The field is then expanded as

$$\mathbf{E}_i = \sum_{m=0}^{\infty}\sum_{n=0}^{\infty} \left(B_{emn}\mathbf{M}_{emn} + B_{omn}\mathbf{M}_{omn} + A_{emn}\mathbf{N}_{emn} + A_{omn}\mathbf{N}_{omn} \right). \tag{A.20}$$

The solution for this expansion is

$$\mathbf{E}_i = E_0 \sum_{n=1}^{\infty} i^n \frac{2n+1}{n(n+1)} \left(\mathbf{M}_{o1n}^{(1)} - i\mathbf{N}_{e1n}^{(1)} \right), \tag{A.21}$$

where the superscript (1) indicates that the radial dependence of $\mathbf{M}$ and $\mathbf{N}$ is given by j_n because of its finiteness at the origin. The corresponding magnetic field is calculated from the curl of (A.21)

$$\mathbf{H}_i = \frac{-k}{\omega\mu} E_0 \sum_{n=1}^{\infty} i^n \frac{2n+1}{n(n+1)} \left(\mathbf{M}_{e1n}^{(1)} + i\mathbf{N}_{o1n}^{(1)} \right). \tag{A.22}$$

The internal and scattered fields

The scattered electromagnetic fields and the fields inside the sphere can also be expanded in vector spherical harmonics. At the boundary between the sphere and the surrounding medium the following conditions have to be fulfilled:

$$(\mathbf{E}_i + \mathbf{E}_s - \mathbf{E}_1) \times \hat{\mathbf{e}}_r = (\mathbf{H}_i + \mathbf{H}_s - \mathbf{H}_1) \times \hat{\mathbf{e}}_r = 0. \qquad (A.23)$$

The solution for the fields inside the sphere is given by

$$\mathbf{E}_1 = \sum_{n=1}^{\infty} E_n \left(c_n \mathbf{M}_{o1n}^{(1)} - i d_n \mathbf{N}_{e1n}^{(1)} \right), \qquad (A.24)$$

$$\mathbf{H}_1 = \frac{-k_1}{\omega \mu_1} \sum_{n=1}^{\infty} E_n \left(d_n \mathbf{M}_{e1n}^{(1)} + i c_n \mathbf{N}_{o1n}^{(1)} \right), \qquad (A.25)$$

where $E_n = i^n E_0 (2n+1)/n(n+1)$, μ_1 is the permeability of the sphere and j_n was again chosen as the appropriate Bessel function due to its behavior at the origin.

The scattered fields result to

$$\mathbf{E}_s = \sum_{n=1}^{\infty} E_n \left(i a_n \mathbf{N}_{e1n}^{(3)} - b_n \mathbf{M}_{o1n}^{(3)} \right), \qquad (A.26)$$

$$\mathbf{H}_s = \frac{k}{\omega \mu} \sum_{n=1}^{\infty} E_n \left(i b_n \mathbf{N}_{o1n}^{(3)} + a_n \mathbf{M}_{e1n}^{(3)} \right), \qquad (A.27)$$

where the superscript (3) indicates that the radial dependence is given by the Hankel functions $h_n^{(1)}$.

Four boundary conditions are implied at $r = a$ by (A.23)

$$E_{i\theta} + E_{s\theta} = E_{1\theta}, \qquad E_{i\phi} + E_{s\phi} = E_{1\phi}, \qquad (A.28)$$

$$H_{i\theta} + H_{s\theta} = H_{1\theta}, \qquad H_{i\phi} + H_{s\phi} = H_{1\phi}, \qquad (A.29)$$

which allow to determine the coefficients a_n, b_n, c_n, and d_n. Substituting the solutions of the electromagnetic fields into this system of linear equations leads to a solution for the scattering coefficients

$$a_n = \frac{\mu m^2 j_n(mx) [x j_n(x)]' - \mu_1 j_n(x) [m x j_n(mx)]'}{\mu m^2 j_n(mx) \left[x h_n^{(1)}(x) \right]' - \mu_1 h_n^{(1)}(x) [m x j_n(mx)]'}, \qquad (A.30)$$

$$b_n = \frac{\mu_1 j_n(mx) [x j_n(x)]' - \mu j_n(x) [m x j_n(mx)]'}{\mu_1 j_n(mx) \left[x h_n^{(1)}(x) \right]' - \mu h_n^{(1)}(x) [m x j_n(mx)]'}, \qquad (A.31)$$

and for the coefficients of the field inside the particle

$$c_n = \frac{\mu_1 j_n(x)\left[xh_n^{(1)}(x)\right]' - \mu_1 h_n^{(1)}(x)\left[xj_n(x)\right]'}{\mu_1 j_n(mx)\left[xh_n^{(1)}(x)\right]' - \mu h_n^{(1)}(x)\left[mxj_n(mx)\right]'}, \quad (A.32)$$

$$d_n = \frac{\mu_1 m j_n(x)\left[xh_n^{(1)}(x)\right]' - \mu_1 m h_n^{(1)}(x)\left[xj_n(x)\right]'}{\mu m^2 j_n(mx)\left[xh_n^{(1)}(x)\right]' - \mu_1 h_n^{(1)}(x)\left[mxj_n(mx)\right]'}, \quad (A.33)$$

where the prime indicates the derivative with respect to the argument in parentheses and the dimensionless size parameter x and the relative refractive index are given by

$$x = ka = \frac{2\pi N a}{\lambda} \quad \text{and} \quad m = \frac{k_1}{k} = \frac{n_s}{n_m}. \quad (A.34)$$

The scattering and extinction cross sections can then be calculated from the scattering coefficients yielding

$$C_{\text{sca}} = \frac{2\pi}{k^2} \sum_{n=1}^{\infty} (2n+1)\left(|a|^2 + |b|^2\right), \quad (A.35)$$

$$C_{\text{ext}} = \frac{2\pi}{k^2} \sum_{n=1}^{\infty} (2n+1)\,\text{Re}\{a_n + b_n\}. \quad (A.36)$$

The absorption cross section can be calculated as

$$C_{\text{abs}} = C_{\text{ext}} - C_{\text{sca}}. \quad (A.37)$$

Numerical evaluation

The numerical evaluation of the cross sections is straight forward. The evaluation of the infinite sum in equations (A.35) and (A.36) needs to be considered carefully. Since the contributions of the subsequent terms become smaller with increasing n, one can chose an appropriate cutoff for the sum. In my evaluation I chose a numerical accuracy of 10^{-5}.

Corrections for small colloid sizes

The bulk complex refractive index of metallic calcium was obtained from experiments [127, 128]. However in the calcium-calcium fluoride system, two corrections need to be taken into account [138]. On the one hand a correction for very small colloid sizes, which is necessary when the colloid size is smaller than the mean free path of the electrons in the material. On the other hand a pressure effect needs to be considered, because of the stress induced in the metallic calcium due to the difference in the lattice constant to the surrounding calcium fluoride matrix.

I will follow the illustration by Orera and Alcalá [138].

Determining the dielectric constant

The dielectric constant for a metal can be written as

$$\epsilon(\omega) = \epsilon_1(\omega) + i\epsilon_2(\omega), \tag{A.38}$$

where

$$\epsilon_1(\omega) = \epsilon_0 - \frac{\omega_p^2}{\omega^2 + \omega_0^2} + B_1(\omega) \tag{A.39}$$

and

$$\epsilon_2(\omega) = \frac{\omega_p^2 \omega_0}{\omega^2 + \omega_0^2} + B_2(\omega). \tag{A.40}$$

ϵ_0 is given by

$$\epsilon_0 = 1 + 4\pi N_I \alpha_I, \tag{A.41}$$

where the atomic concentration N_I is given by

$$N_I = \frac{N_A \cdot \rho_{Ca}}{M_{Ca}} = 2.33 \cdot 10^{-2} \text{Å}^{-3} \tag{A.42}$$

and the experimental value of the electronic polarizability of calcium is $\alpha_I(\text{Ca}^{2+}) = 1.1\,\text{Å}^3$ [152]. The plasma frequency ω_p can be written as

$$\omega_p^2 = \frac{4\pi n_0 e^2}{m_{\text{opt}}}, \tag{A.43}$$

where n_0 is the electron concentration and m_{opt} the optical electron mass. I will use $\hbar\omega_p = 5.78\,\text{eV}$ [138]. ω_0 is the electron collision frequency and is given by $[\omega_0]_{\text{bulk}} = 0.27 \cdot 10^{14} s^{-1}$ for bulk calcium [138]. The inter band contributions are obtained from the experimental values of the dielectric constant

$$B_1(\omega) = \epsilon_1(\exp) - \epsilon_0 + \frac{\omega_p^2}{\omega^2 + [\omega_0]_{\text{bulk}}^2} \tag{A.44}$$

$$B_2(\omega) = \epsilon_2(\exp) - \frac{\omega_p^2 [\omega_0]_{\text{bulk}}}{\omega^2 + [\omega_0]_{\text{bulk}}^2}. \tag{A.45}$$

Mean free path correction

The mean free path correction is introduced to include an increase of the collision frequency, which occurs because the particle dimensions are smaller than the mean free path of the electrons because the electrons scatter at the particle boundaries. The corrected collision frequency

is given by
$$\omega_0 = [\omega_0]_{\text{bulk}} + \frac{v_F}{a}, \tag{A.46}$$
where v_F is the Fermi velocity of the electrons and a is the particle radius. The Fermi velocity v_F can be obtained from
$$v_F = \sqrt{\frac{2E_F}{m_e}} = \frac{\hbar}{m_e}\left(\frac{3\pi^2 Z_v \rho N_A}{M_{\text{Ca}}}\right)^{1/3} = 1.288 \cdot 10^6 \frac{m}{s}, \tag{A.47}$$
where Z_v is number of valence electrons per atom.

Pressure effect

Due to the difference of the lattice constants of calcium and calcium fluoride, the calcium colloids are compressed by the CaF$_2$ matrix. The resulting pressure effect on ω_0 is negligible, while the corrections to $B_1(\omega)$ and $B_2(\omega)$ are below 1% and 5% respectively [138]. The plasma frequency is corrected by an empiric factor $\omega_p(0)/\omega_p(P) = 1.11$, where $\omega_p(P)$ is the plasma frequency at pressure P [138].

Appendix B

Electronic structure results for CaF_2 and its defect structures

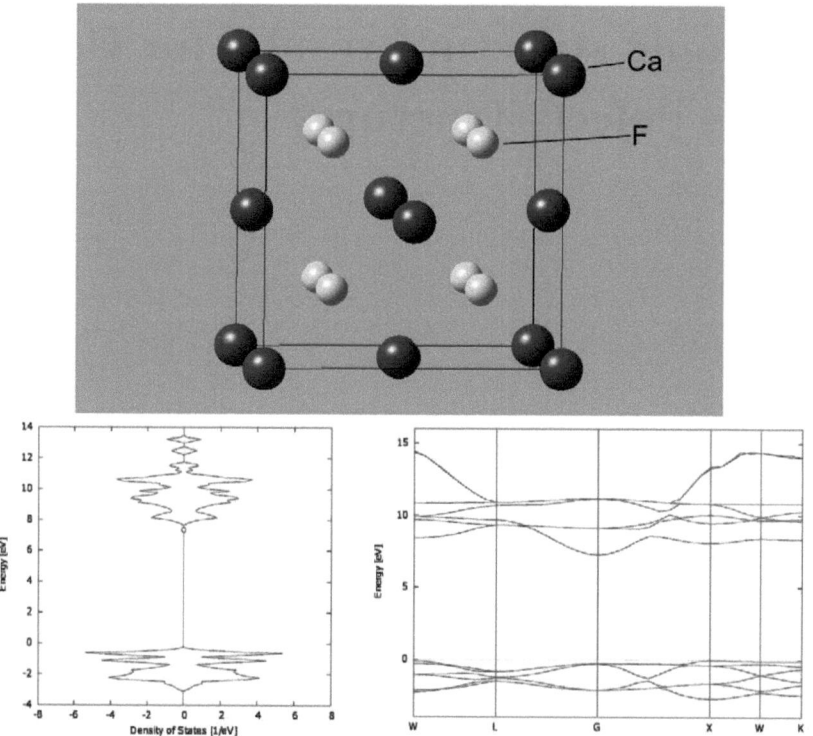

Figure B.1: Structure optimized configuration (top), density of states (bottom left) and band structure (bottom right) of CaF_2.

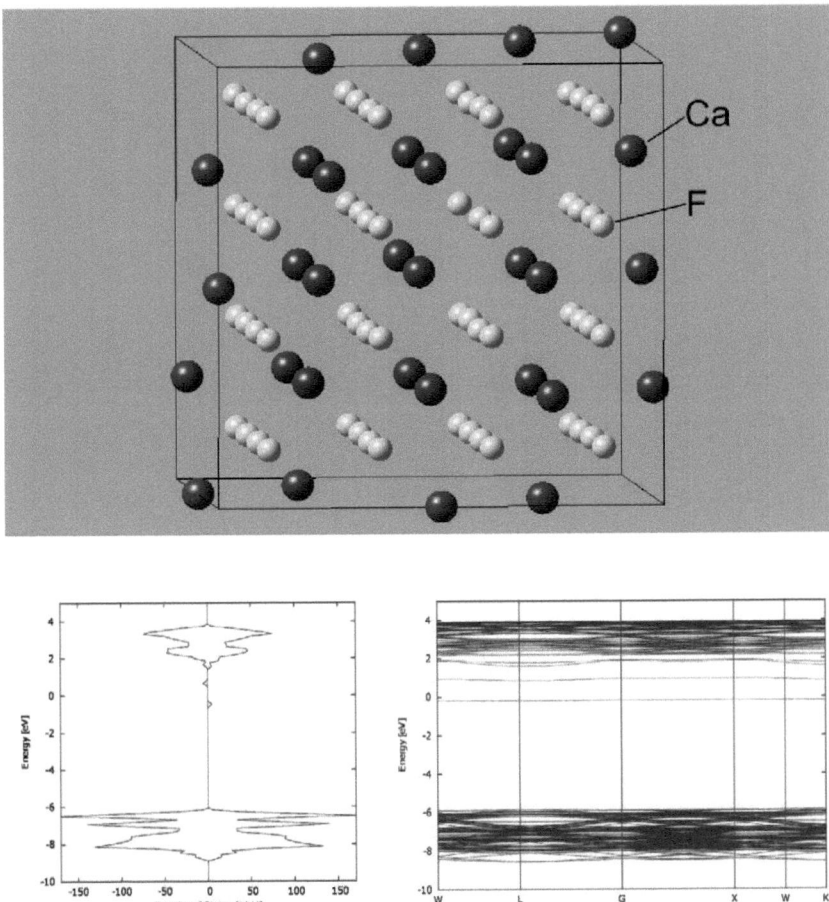

Figure B.2: Structure optimized configuration (top), density of states (bottom left) and band structure (bottom right) of the F center in CaF_2.

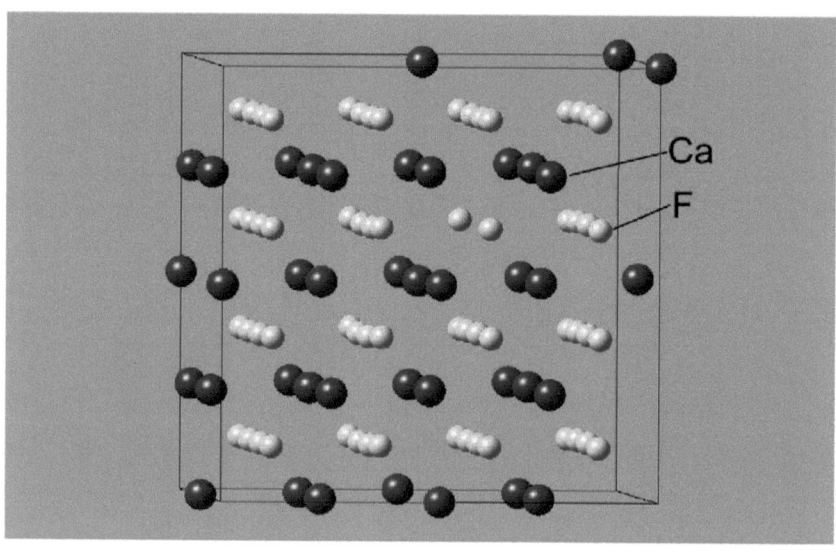
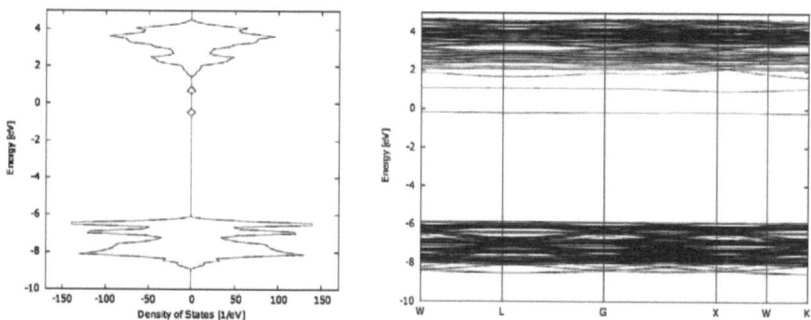

Figure B.3: Structure optimized configuration (top), density of states (bottom left) and band structure (bottom right) of the M center in CaF_2.

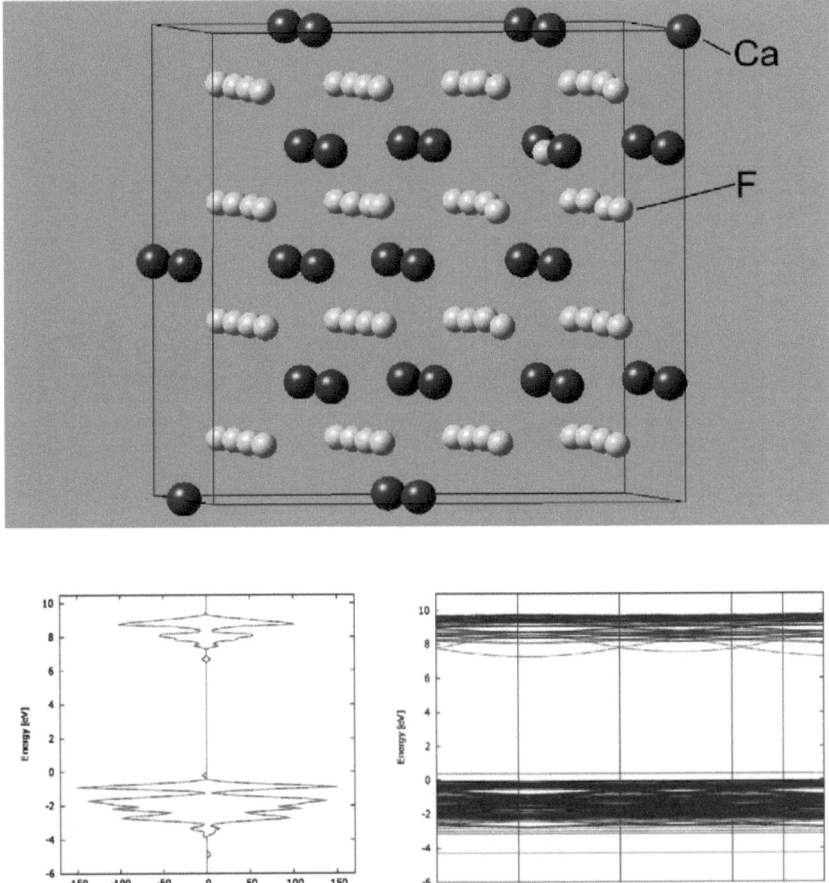

Figure B.4: Structure optimized configuration (top), density of states (bottom left) and band structure (bottom right) of the H center in CaF_2.

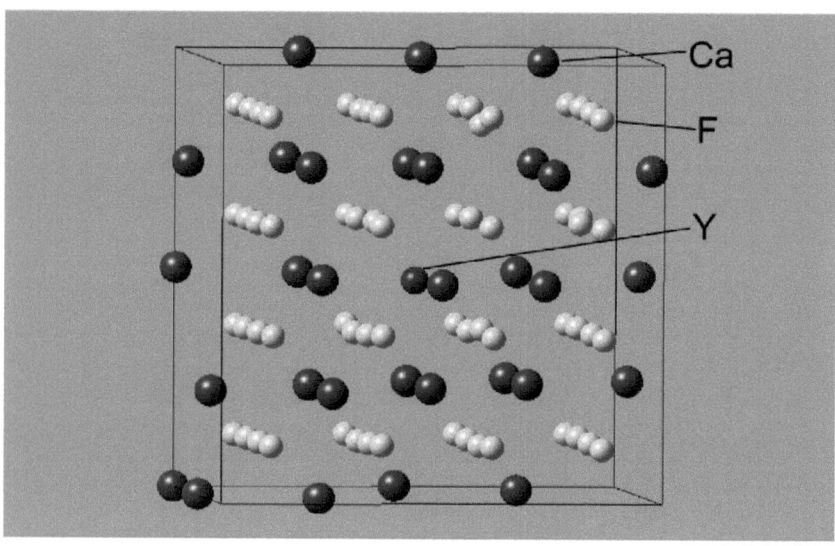
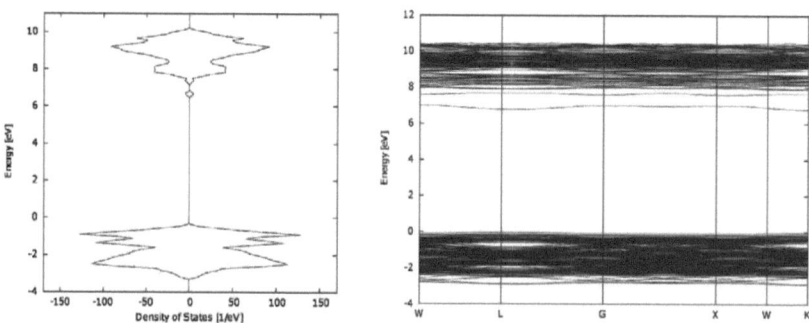

Figure B.5: Structure optimized configuration (top), density of states (bottom left) and band structure (bottom right) of the F_{Na} center in CaF_2.

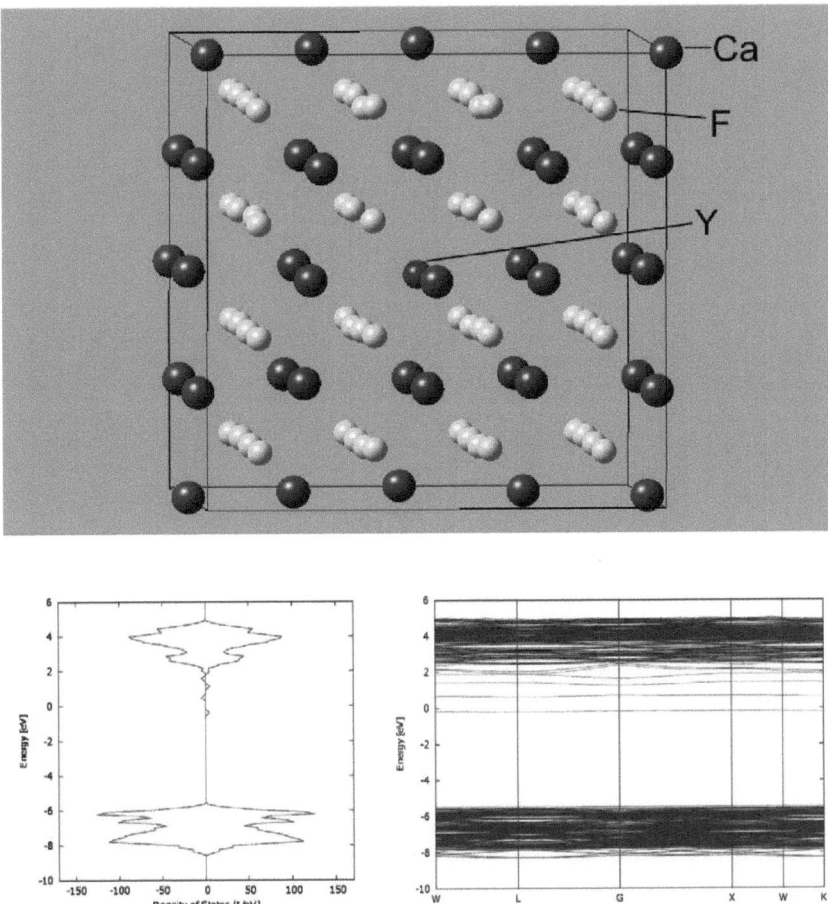

Figure B.6: Structure optimized configuration (top), density of states (bottom left) and band structure (bottom right) of the M_{Na} center in CaF_2.

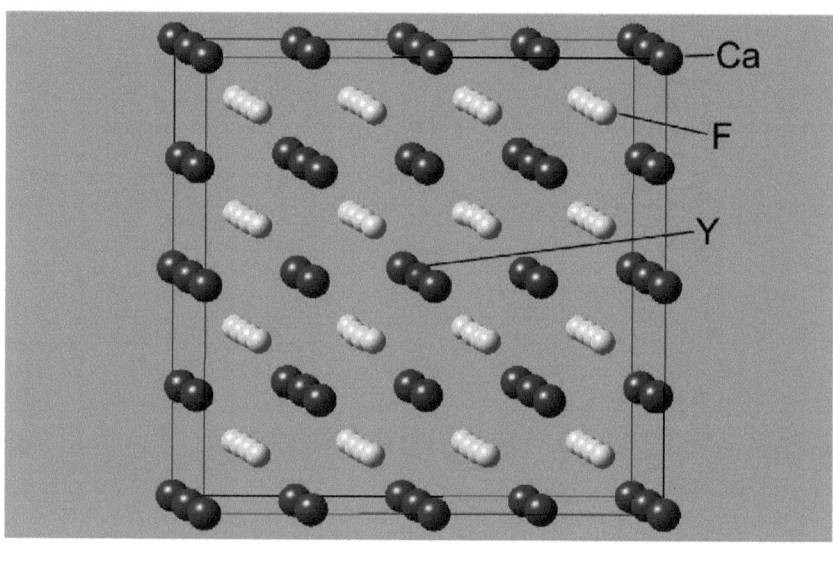

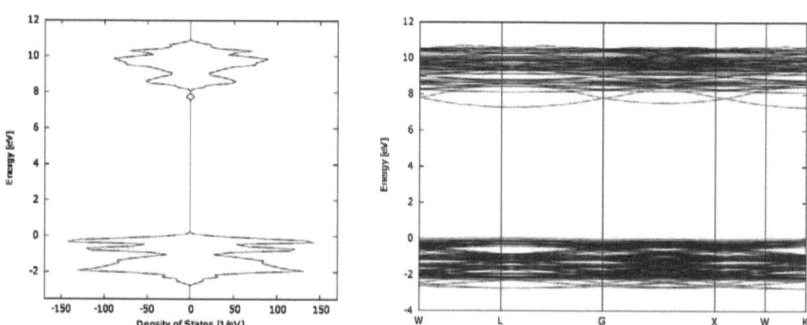

Figure B.7: Structure optimized configuration (top), density of states (bottom left) and band structure (bottom right) of a Na$^+$ impurity in CaF$_2$.

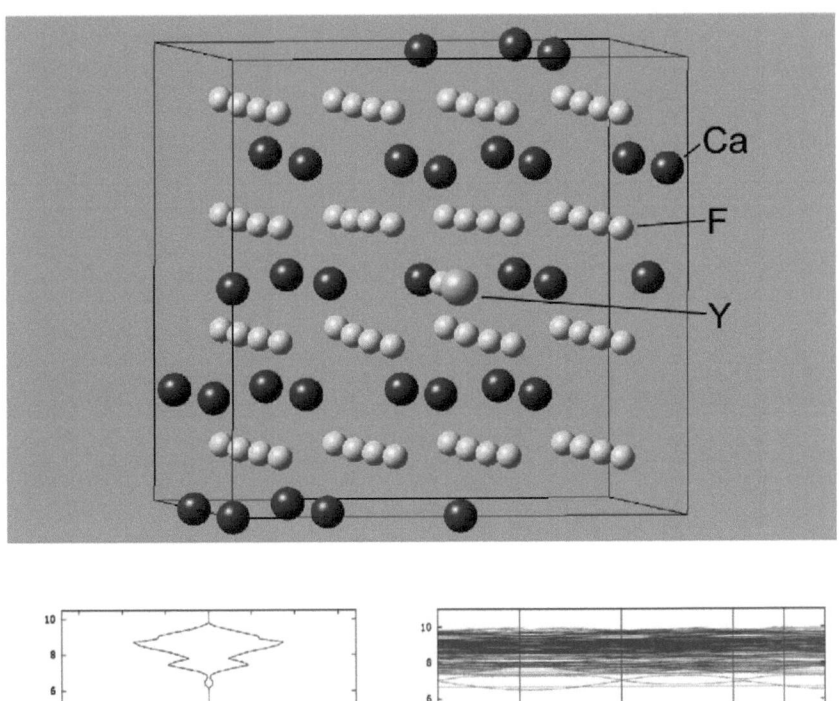

Figure B.8: Structure optimized configuration (top), density of states (bottom left) and band structure (bottom right) of the H_Y center in CaF_2.

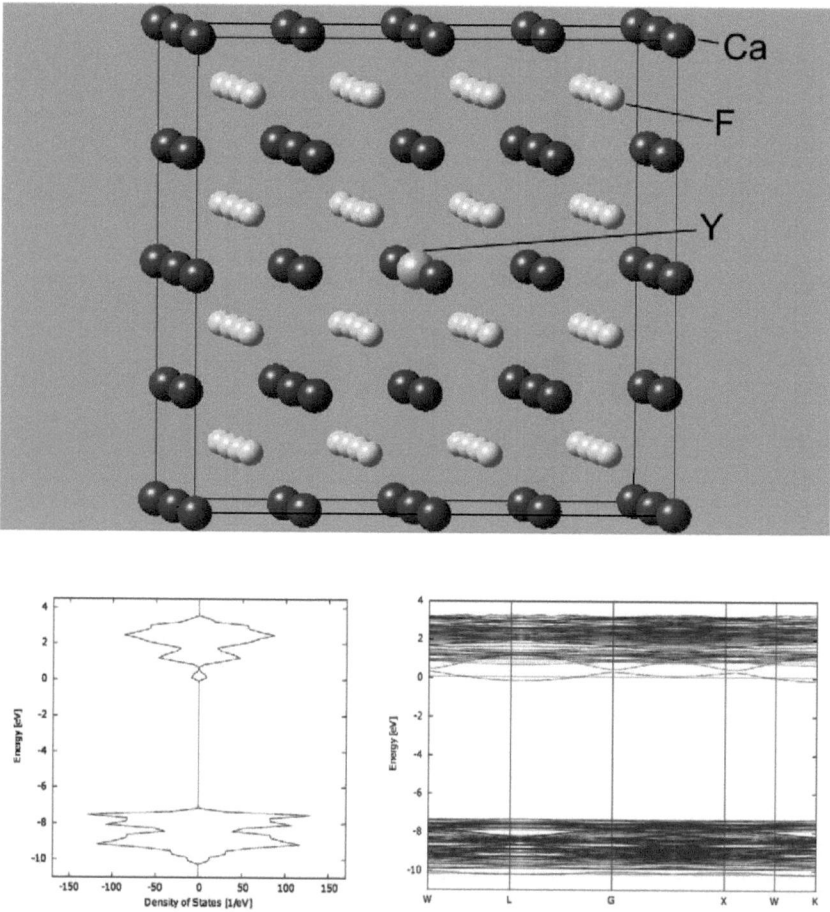

Figure B.9: Structure optimized configuration (top), density of states (bottom left) and band structure (bottom right) of an Y^{3+} impurity in CaF_2.

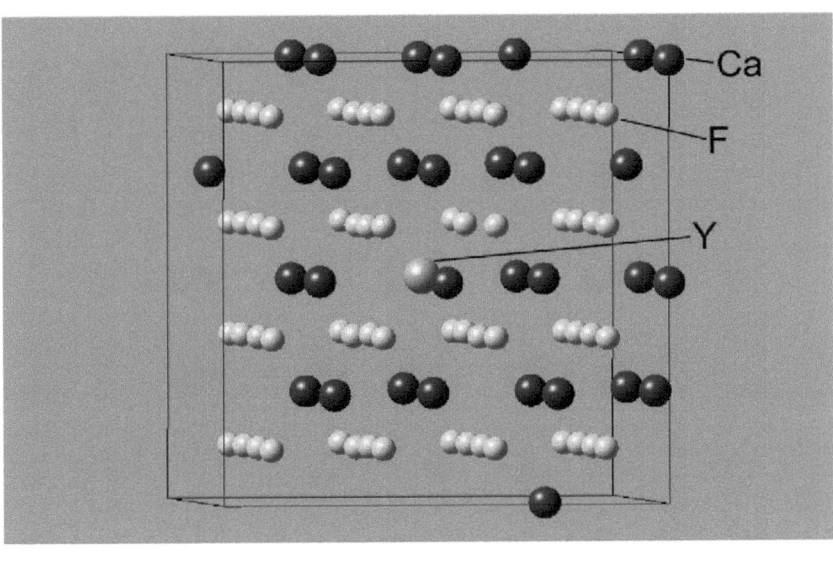

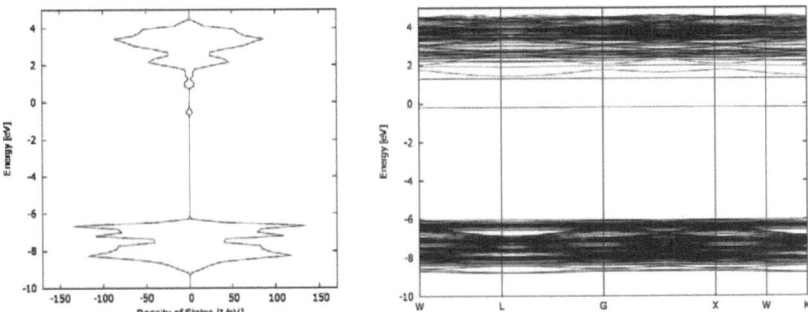

Figure B.10: Structure optimized configuration (top), density of states (bottom left) and band structure (bottom right) of the F_Y center in CaF_2.

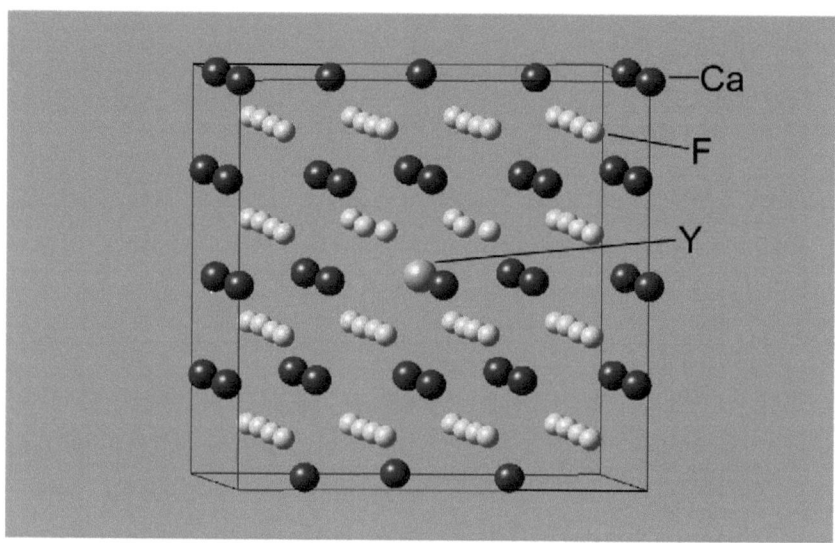

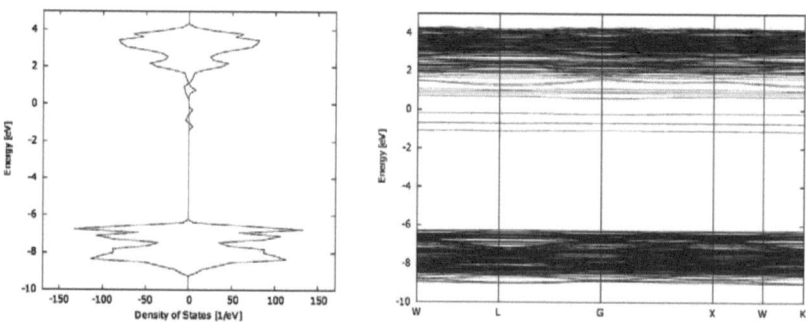

Figure B.11: Structure optimized configuration (top), density of states (bottom left) and band structure (bottom right) of the M_Y center in CaF_2.

Bibliography

[1] O. Schott. Ueber Glasschmelzerei für optische und andere wissenschaftliche Zwecke. In *Verhandlungen des Vereins zur Beförderung des Gewerbefleisses vom 4. Juni 1888*, 162, (1888).

[2] E. Abbe. *Abhandlungen über die Theorie des Mikroskops, Vol. 1 der Gesammelten Abhandlungen*, pp. 478–486. Georg Olms Verlag, (1989).

[3] M. Zajac and J. Nowak. Correction of chromatic aberration in hybrid objectives. *Optik - International Journal for Light and Electron Optics*, **113**, 299, (2002).

[4] K. Striegel. Close to the action. *Schott Solutions*, **1**, 38, (2008).

[5] J. Lucas, F. Smektala, and J. L. Adam. Fluorine in optics. *J. Fluorine Chem.*, **114**, 113, (2002).

[6] Hellma Materials GmbH & Co. KG. Data Sheet Calcium Fluoride. www.hellma-materials.com/text/966/en/produktinformation.html, (accessed December 2010).

[7] P. Hinsmann, J. Frank, P. Svasek, M. Harasek, and B. Lendl. Design, simulation and application of a new micromixing device for time resolved infrared spectroscopy of chemical reactions in solution. *Lab on a Chip*, **1**, 16, (2001).

[8] T. Pan, R. T. Kelly, M. C. Asplund, and A. T. Woolley. Fabrication of calcium fluoride capillary electrophoresis microdevices for on-chip infrared detection. *J. Chromatography A*, **1027**, 231, (2004).

[9] G. E. Moore. Cramming more components onto integrated circuits. *Electronics*, **38**, 114, (1965).

[10] M. Rothschild. Projection optical lithography. *Materials Today*, **8**, 18, (2005).

[11] W. Kaiser and P. Kuerz. EUVL - Extreme Ultraviolet Lithography. *Optik und Photonik*, **2**, 35, (2009).

[12] J. H. Burnett, Z. H. Levine, and E. L. Shirley. Intrinsic birefringence in calcium fluoride and barium fluoride. *Phys. Rev. B*, **64**, 241102, (2001).

[13] M. Letz, L. Parthier, A. Gottwald, and M. Richter. Spatial anisotropy of the exciton level in CaF_2 at 11.1 eV and its relation to the weak optical anisotropy at 157 nm. *Phys. Rev. B*, **67**, 233101, (2003).

[14] C. Muehlig, W. Triebel, G. Toepfer, J. Bergmann, S. Brueckner, C. Chojetzki, and R. Martin. Laser induced fluorescence of calcium fluoride upon 193 nm and 157 nm excitation. *Proc. SPIE*, **5188**, 123, (2003).

[15] U. Natura, S. Rix, M. Letz, and L. Parthier. Study of haze in 193nm high dose irradiated CaF_2 crystals. *Proc. SPIE*, **7504**, 75041P, (2009).

[16] S. Rix, U. Natura, M. Letz, C. Felser, and L. Parthier. A microscopic model for long-term laser damage in calcium fluoride. *Proc. SPIE*, **7504**, 75040J, (2009).

[17] C. Muehlig, W. Triebel, H. Stafast, and M. Letz. Influence of Na-related defects on ArF laser absorption in CaF_2. *Appl. Phys. B*, **99**, 525, (2010).

[18] C. Doelter. Über die Farben der Mineralien. *Die Naturwissenschaften*, **8**, 4, (1920).

[19] C. Doelter. Beobachtung über Verfärbung von Mineralen durch Bestrahlung. *Tschermaks Mineralogische und Petrographische Mitteilungen*, **38**, 456, (1925).

[20] C. Doelter. Bedeutung der Kolloidchemie für Mineralogie und Geologie. *Kolloidchemische Zeitschrift, Zsigmondy-Festsschrift*, **Erg.-Bd. 36**, 95, (1925).

[21] G. Mie. Beiträge zur Optik trüber Medien, speziell kolloidaler Metallösungen. *Annalen der Physik*, **330**, 377, (1908).

[22] L. Göbel. Radioaktive Zersetzungserscheinungen am Flussspat. *Zeitschrift für Kristallographie*, **76**, 457, (1931).

[23] K. Przibram. Verfärbung und Lumineszenz durch Bequerelstrahlung. *Zeitschrift für Physik A*, **102**, 331, (1936).

[24] K. Przibram. Fluorescence of the Bivalent Rare Earths. *Nature*, **139**, 329, (1937).

[25] K. Przibram. Absorption Bands and Electron Transitions in Coloured Fluorites. *Nature*, **141**, 970, (1938).

[26] H. Bill and R. Lacroix. Preliminary results on a centre in CaF_2 crystals. *Phys. Lett.*, **21**, 257, (1966).

[27] H. Bill and R. Lacroix. EPR of a centre in Y^{3+} doped artificial CaF_2 crystals. *Phys. Lett.*, **22**, 250, (1966).

[28] H. Bill and R. H. Silsbee. Dynamical Jahn-Teller and reorientation effects in the EPR spectrum of $CaF_2:O^-$. *Phys. Rev. B*, **10**, 2697, (1974).

[29] H. Bill and G. Calas. Color Centers, Asssociated Rare-Earth Ions and the Origin of Coloration in Natural Fluorites. *Phys. Chem. Min.*, **3**, 117, (1978).

[30] H. Bill. Investigation of colour centres in alkaline earth fluorides. *Helv. Phys. Acta*, **42**, 771, (1969).

[31] H. Bill. ENDOR investigation of Gd^{3+} in CaF_2 crystals. *Phys. Lett. A*, **29**, 593, (1969).

[32] H. Bill and J. Mareda. ENDOR investigation of a V_F center in natural CaF_2 crystals. *Chem. Phys. Lett.*, **36**, 218 , (1975).

[33] W. B. Fowler. *Physics of Color Centers*. Academic Press, (1968).

[34] W. Hayes. *Crystals with the fluorite structure*. Oxford University Press, (1974).

[35] W. Gellermann. Color center lasers. *J. Phys. Chem. Solids*, **1**, 249, (1991).

[36] J. M. G. Tijero and F. Jaque. Thermal and optical properties of the F_A and $(F_2^+)_A$ centers in Na-doped CaF_2 crystals. *Phys. Rev. B*, **41**, 3832, (1990).

[37] A. Monnier, M. Schnieper, R. Jaaniso, and H. Bill. Samarium doped alkaline earth halide thin films as spectrally selective materials for hole burning. *Radiation Effects and Defects in Solids*, **135**, 253, (1995).

[38] H. Matzke. CaF_2 als Modellsubstanz für Diffusionsvorgänge in UO_2. *J. Nuc. Mat.*, **11**, 344, (1964).

[39] A. Pryor. Thermal vibration amplitudes in fluorite lattices. *J. Phys. Chem. Sol.*, **26**, 2045, (1965).

[40] M. Nazmy and M. Abdel-Azim. Observations on the radial and axial shrinkage behaviour of CaF_2 and UO_2 Powder compacts. *J. Nuc. Mat.*, **52**, 296, (1974).

[41] P. J. D. Lindan and M. J. Gillan. A molecular dynamics study of the thermal conductivity of CaF_2 and UO_2. *J. Phys. Cond. Mat.*, **3**, 3929, (1991).

[42] V. F. Shtan'ko and E. P. Chinkov. Time-resoved spectroscopy of self-trapped excitons in fluorides of alkaline-earth metals under pulsed electron irradiation. *Phys. Sol. Stat.*, **40**, 1119, (1998).

[43] V. Denks, A. Maaroos, V. Nagirnyi, T. Savikhina, and V. Vassiltsenko. Excitonic processes in Li- and Na-doped CaF_2. *Radiation Effects and Defects in Solids*, **149**, 297, (1999).

[44] V. Denks, A. Maaroos, V. Nagirnyi, T. Savikhina, and V. Vassiltsenko. Excitonic processes in pure and doped CaF_2. *J. Phys. Cond. Mat.*, **11**, 3115, (1999).

[45] A. Franceschetti and A. Zunger. The inverse band-structure problem of finding an atomic configuration with given electronic properties. *Nature*, **402**, 60, (1999).

[46] E. Wimmer. Prediction of Materials Properties. In P. von Ragué-Schleyer, editor, *Encyclopedia of Computational Chemistry*. Wiley, (1998).

[47] R. Car and M. Parrinello. Unified Approach for Molecular Dynamics and Density-Functional Theory. *Phys. Rev. Lett.*, **55**, 2471, (1985).

[48] P. A. M. Dirac. Note on the Exchange Phenomena in the Thomas Atom. *Proc. Cambridge Phil. Roy. Soc.*, **26**, 376, (1930).

[49] D. R. Hartree. The Wave Mechanics of an Atom with a Non-Coulomb Central Field. Part I. Theory and Methods. *Proc. Cambridge Phil. Soc.*, **24**, 89, (1928).

[50] V. Fock. Näherungsmethode zur Lösung des quantenmechanischen Mehrkörperproblems. *Zeitschrift für Physik*, **61**, 126, (1930).

[51] P. Hohenberg and W. Kohn. Inhomogeneous Electron Gas. *Phys. Rev.*, **136**, B864, (1964).

[52] W. Kohn and L. J. Sham. Self-Consistent Equations Including Exchange and Correlation Effects. *Phys. Rev. A*, **140**, 1133, (1965).

[53] J. Anderson. Quantum Monte Carlo: Atoms, molecules, clusters, liquids, and solids. *Rev. Comp. Chem.*, **13**, 133, (1999).

[54] K. Binder and D. Heermann. *Monte Carlo simulation in statistical physics: an introduction*. Springer-Verlag, (2010).

[55] L. H. Thomas. The calculation of atomic fields. *Proc. Cambridge Phil. Roy. Soc.*, **23**, 542, (1927).

[56] E. Fermi. Un metodo statistico per la determinazione di alcune priorieta dell'atome. *Rend. Accad. Naz. Lincei*, **6**, 602, (1927).

BIBLIOGRAPHY

[57] E. Fermi. Eine statistische Methode zur Bestimmung einiger Eigenschaften des Atoms und ihre Anwendung auf die Theorie des periodischen Systems der Elemente. *Zeitschrift für Physik A*, **48**, 73, (1928).

[58] A. D. Becke. Density-functional exchange-energy approximation with correct asymptotic behavior. *Phys. Rev. A*, **38**, 3098, (1988).

[59] J. P. Perdew and Y. Wang. Accurate and simple analytic representation of the electron-gas correlation energy. *Phys. Rev. B*, **45**, 13244, (1992).

[60] J. P. Perdew, K. Burke, and M. Ernzerhof. Generalized Gradient Approximation Made Simple. *Phys. Rev. Lett.*, **77**, 3865, (1996).

[61] V. Anisimov, F. Aryasetiawan, and A. Lichtenstein. First-principles calculations of the electronic structure and spectra of strongly correlated systems: the LDA+ U method. *J. Phys. Cond. Matt.*, **9**, 767, (1997).

[62] C. Lee, W. Yang, and R. G. Parr. Development of the Colle-Salvetti correlation-energy formula into a functional of the electron density. *Phys. Rev. B*, **37**, 785, (1988).

[63] L. N. Oliveira, E. K. U. Gross, and W. Kohn. Density-Functional Theory for Superconductors. *Phys. Rev. Lett.*, **60**, 2430, (1988).

[64] R. M. Martin. *Electronic Structure*. Cambridge University Press, (2008).

[65] C. Fiolhais, F. Nogueira, and M. Marques, editors. *A Primer in Density Functional Theory*. Springer-Verlag, (2003).

[66] M. Born and R. Oppenheimer. Zur Quantentheorie der Molekeln. *Annalen der Physik*, **389**, 457, (1927).

[67] H. A. Jahn and E. Teller. Stability of polyatomic molecules in degenerate electronic states. I. Orbital degeneracy. *Proc. Roy. Soc. London A*, **161**, 220, (1937).

[68] H. Fröhlich. Theory of the superconducting state. I. The ground state at the absolute zero of temperature. *Phys. Rev.*, **79**, 845, (1950).

[69] F. Herman, J. P. Van Dyke, and I. B. Ortenburger. Improved Statistical Exchange Approximation for Inhomogeneous Many-Electron Systems. *Phys. Rev. Lett.*, **22**, 807, (1969).

[70] D. Vanderbilt. Soft self-consistent pseudopotentials in a generalized eigenvalue formalism. *Phys. Rev. B*, **41**, 7892, (1990).

[71] P. E. Blöchl. Projector augmented-wave method. *Phys. Rev. B*, **50**, 17953, (1994).

[72] G. Kresse and D. Joubert. From ultrasoft pseudopotentials to the projector augmented-wave method. *Phys. Rev. B*, **59**, 1758, (1999).

[73] C. Wert and C. Zener. Interstitial Atomic Diffusion Coefficients. *Phys. Rev.*, **76**, 1169, (1949).

[74] H. Eyring. The Activated Complex in Chemical Reactions. *J. Chem. Phys.*, **3**, 107, (1935).

[75] C.Dellago, P.G.Bolhuis, and P. Geissler. Transition Path Sampling. *Adv. Chem. Phys.*, **123**, 1, (2002).

[76] G. Henkelmann, B. P. Uberuaga, and H. Jonsson. A climbing image nudged elastic band method for finding saddle points and minimum energy paths. *J. Chem. Phys.*, **113**, 9901, (2000).

[77] E. Wimmer, W. Wolf, J. Sticht, P. Saxe, C. B. Geller, R. Najafabadi, and G. A. Young. Temperature-dependent diffusion coefficients from ab initio computations: Hydrogen, deuterium, and tritium in nickel. *Phys. Rev. B*, **77**, 134305, (2008).

[78] G. H. Vineyard. Frequency factors and isotope effects in solid state rate processes. *J. Phys. Chem. Solids*, **3**, 121, (1957).

[79] R. M. Martin. *Electronic Structure*, pp. 387–405. Cambridge University Press, (2008).

[80] S. Baroni, S. de Gironcoli, A. Dal Corso, and P. Giannozzi. Phonons and related crystal properties from density-functional perturbation theory. *Rev. Mod. Phys.*, **73**, 515, (2001).

[81] H. Hellmann. *Einführung in die Quantenchemie*. F. Deuticke Leipzig, (1937).

[82] R. P. Feynman. Forces in Molecules. *Phys. Rev.*, **56**, 340, (1939).

[83] P. Giannozzi, S. Baroni, N. Bonini, M. Calandra, R. Car, C. Cavazzoni, D. Ceresoli, G. L. Chiarotti, M. Cococcioni, I. Dabo, A. Dal Corso, S. de Gironcoli, S. Fabris, G. Fratesi, R. Gebauer, U. Gerstmann, C. Gougoussis, A. Kokalj, M. Lazzeri, L. Martin-Samos, N. Marzari, F. Mauri, R. Mazzarello, S. Paolini, A. Pasquarello, L. Paulatto, C. Sbraccia, S. Scandolo, G. Sclauzero, A. P. Seitsonen, A. Smogunov, P. Umari, and R. M. Wentzcovitch. QUANTUM ESPRESSO: a modular and open-source software project for quantum simulations of materials. *J. Phys. Cond. Matt.*, **21**, 395502 (19pp), (2009).

[84] D. Alfè. PHON: A program to calculate phonons using the small displacement method. *Comp. Phys. Comm.*, **180**, (2009).

[85] Crystran Ltd. Calcium Fluoride (CaF$_2$) Data Sheet. http://www.crystran.co.uk/uploads/files/96.pdf, (accessed Jan 2011).

[86] A. D. Becke and K. E. Edgecombe. A simple measure of electron localization in atomic and molecular systems. *J. Chem. Phys.*, **92**, 5397, (1990).

[87] J. Arends. Color Centers in Additively Colored CaF$_2$ and BaF$_2$. *Phys. Stat. Sol.*, **7**, 805, (1964).

[88] T. Kamikawa, Y. Kazumata, A. Kikuchi, and L. Ozawa. The F center in calcium fluoride. *Phys. Lett.*, **21**, 126, (1966).

[89] B. Cavenett, W. Hayes, I. Hunter, and A. Stoneham. Magneto optical properties of F centres in alkaline earth fluorides. *Proc. Roy. Soc. London A*, **309**, 53, (1969).

[90] W. Hayes. Point defects in alkaline earth fluorides. *Radiation Effects and Defects in Solids*, **4**, 239, (1970).

[91] Y. Ma and M. Rohlfing. Optical excitation of deep defect levels in insulators within many-body perturbation theory: The F center in calcium fluoride. *Phys. Rev. B*, **77**, 115118, (2008).

[92] J. B. Beaumont and W. Hayes. M centres in alkaline earth fluorides. *Proc. Roy. Soc. A*, **309**, 41, (1969).

[93] R. Rauch and G. Schwotzer. Disturbed Colour Centres in Oxygen- and Alkali-Doped Alkaline Earth Fluoride Crystals after X-Ray Irradiation at 77 and 295 K. *Phys. Stat. Sol.*, **74**, 123, (1982).

[94] C. Muehlig, W. Triebel, G. Toepfer, and A. Jordanov. Calcium fluoride for ArF laser lithography: characterization by in-situ transmission and LIF measurements. *Proc. SPIE*, **4932**, 458, (2003).

[95] A. Burkert, C. Muehlig, W. Triebel, D. Keutel, U. Natura, L. Parthier, S. Gliech, S. Schroeder, and A. Duparre. Investigating the ArF laser stability of CaF$_2$ at elevated fluences. *Proc. SPIE*, **5878**, 58780E, (2005).

[96] C. Mühlig, H. Stafast, W. Triebel, T. Zeuner, C. Karras, and M. Letz. Influence of Na-related defects on DUV nonlinear absorption in CaF$_2$: nanosecond versus femtosecond laser pulses. *Proc. SPIE*, **7504**, 75040I, (2009).

[97] M. Norgett and A. Stoneham. The self trapped hole in alkaline earth fluorides: I. Static properties. *J. Phys. C*, **6**, 229, (1973).

[98] S. Parker, K. Song, C. Catlow, and A. Stoneham. Geometry and charge distribution of H centres in the fluorite structure. *J. Phys. C*, **14**, 4009, (1981).

[99] M. J. Norgett and A. M. Stoneham. The self trapped hole in alkaline earth fluorides: II. Hopping motion. *J. Phys. C*, **6**, 238, (1973).

[100] K. Atobe. Thermoluminescence and F-center annealing in alkaline-earth fluoride crystals after reactor irradiation at low temperature. *J. Chem. Phys.*, **71**, 2588, (1979).

[101] W. Hayes, D. L. Kirk, and G. P. Summers. The self-trapped hole and the self-trapped exciton in alkaline earth fluorides. *Sol. Stat. Commun.*, **7**, 1061, (1969).

[102] R. T. Williams, M. N. Kabler, W. Hayes, and J. P. Stott. Time-resolved spectroscopy of self-trapped excitons in fluorite crystals. *Phys. Rev. B*, **14**, 725, (1976).

[103] M. N. Kabler and R. T. Williams. Vacancy-interstitial pair production via electron-hole recombination in halide crystals. *Phys. Rev. B*, **18**, 1948, (1978).

[104] M. Adair, C. Leung, and K. Song. Equilibrium configuration of the self-trapped exciton in CaF2 and SrF2. *J. Phys. C*, **18**, L909, (1985).

[105] K. Song, C. Leung, and J. Spaeth. Zero-field splitting of the self-trapped exciton in alkali fluorides and alkaline-earth fluorides. *J. Phys.*, **2**, 6373, (1990).

[106] N. Itoh and K. Tanimura. Formation of inerstitial-vacancy pairs by electronic excitation in pure ionic crystals. *J. Phys. Chem. Solids*, **51**, 717, (1990).

[107] K. Tanimura. Femtosecond time-resolved spectroscopy of the formation of self-trapped excitons in CaF_2. *Phys. Rev. B*, **63**, 184303, (2001).

[108] C. Görling, U. Leinhos, and K. Mann. Self-trapped exciton luminescence and repetition rate dependence of two-photon absorption in CaF_2 at 193 nm. *Opt. Commun.*, **216**, 369, (2003).

[109] L. P. Cramer, T. D. Cumby, J. A. Leraas, S. C. Langford, and J. T. Dickinson. Effect of surface treatments on self-trapped exciton luminescence in single-crystal CaF_2. *J.Appl. Phys.*, **97**, 103533, (2005).

[110] A. Franklin. Born model calculation of defect energies in CaF2. *J. Phys. Chem. Sol.*, **29**, 823, (1968).

[111] H. Bill and W. von der Osten. Raman and Vibronic Spectra of a New Oxygen Molecule Ion in CaF2. *Phys. Stat. Sol. B*, **75**, 613, (1976).

[112] J. Weber and H. Bill. Jahn-Teller distortion and electronic structure of the O-center in CaF2: an MS X α study. *Chem. Phys. Lett.*, **52**, 562, (1977).

[113] W. J. Scouler and A. Smakula. Coloration of Pure and Doped Calcium Fluoride Crystals at 20°C and -190°C. *Phys. Rev.*, **120**, 1154, (1960).

[114] P. D. Southgate. Anelastic and dielectric loss in yttrium-doped calcium fluoride. *Journal of Physics and Chemistry of Solids*, **27**, 1623, (1966).

[115] G. von der Gönna, L. Parthier, G. Wehrhan, and M. Letz. CaF_2-Einkristalle mit erhöhter Laserstabilität, Verfahren zu ihrer Herstellung und ihre Verwendung. DE Patent 10 2005 044 697 A1, (2007).

[116] A. V. Puchina, V. E. Puchin, E. A. Kotomin, and M. Reichling. Ab initio study of the F centers in CaF_2: Calculations of the optical absorption, diffusion and binding energies. *Sol. Stat. Comm.*, **106**, 285, (1998).

[117] R. Wehn and P. Lunkenheimer. Bestimmung der dielektrischen Eigenschaften von CaF_2 Proben (Auftragsmessung), (2007).

[118] M. Letz and L. Parthier. Charge centers in CaF_2: Ab initio calculation of elementary physical properties. *Phys. Rev. B*, **74**, 064116, (2006).

[119] R. W. Ure, Jr. Ionic Conductivity of Calcium Fluoride Crystals. *J. Chem. Phys.*, **26**, 1363, (1957).

[120] G. A. Keig and R. L. Coble. Mobility of Edge Dislocations in Single-Crystal Calcium Fluoride. *J. Appl. Phys.*, **39**, 6090, (1968).

[121] M. Huisinga, M. Reichling, and E. Matthias. Ultraviolet photoelectron spectroscopy and photoconductivity of CaF_2. *Phys. Rev. B*, **55**, 7600, (1996).

[122] S. C. Keeton and W. D. Wilson. Vacancies, Interstitials, and Rare Gases in Fluorite Structures. *Phys. Rev. B*, **7**, 834, (1973).

[123] D. Chakravorty. Energy of migration of anion vacancy and interstitial in calcium fluoride. *J. Phys. Chem. Sol.*, **32**, 1091, (1971).

[124] H. Matzke. Fluorine Self-Diffusion in CaF_2 and BaF_2. *J. Mat. Sci.*, **5**, 831, (1970).

[125] J. W. Twidell. Radiation induced movement of charge compensating ions in CaF_2. *J. Phys. Chem. Sol.*, **31**, 299, (1970).

[126] Corning Inc. Calcium Fluoride CaF$_2$, Physical and Chemical Properties. http://www.corning.com/docs/specialtymaterials/pisheets/ H0607_CaF2_Product_Sheet.pdf, (accessed January 2011).

[127] A. G. Mathewson and H. P. Myers. Absolute Values of the Optical Constants of Some Pure Metals. *Phys. Scr.*, **4**, 291, (1971).

[128] P. O. Nilsson and G. Forssell. Optical properties of calcium. *Phys. Rev. B*, **16**, 3352, (1977).

[129] M. M. Choy, W. R. Cook, R. F. S. Hearmon, H. Jaffe, J. Jerphagnon, S. K. Kurtz, S. T. Liu, and D. F. Nelson. *Landolt-Börnstein, Zahlenwerte und Funktionen aus Naturwissenschaft und Technik*, Volume 11, pp. 9,26. Springer-Verlag, (1979).

[130] J. F. Nye. *Physical Properties of Crystals*, pp. 134–135. Oxford University Press, (1985).

[131] W. Ostwald. Über die vermeintliche Isomerie des roten und gelben Quecksilberoxyds und die Oberflächenspannung fester Körper. *Z. Phys. Chem.*, **6**, 495, (1900).

[132] E. Abbe. *Abhandlungen über die Theorie des Mikroskops, Vol. 1 der Gesammelten Abhandlungen*, pp. 45–100. Georg Olms Verlag, (1989).

[133] U. Natura. Internal report (Schott AG), (2008).

[134] M. Reichling. Bericht über hochauflösende Messungen mit dynamischer Kraftmikroskopie an mit Laserlicht bestrahlten CaF$_2$-Proben von Schott Lithotec (Auftragsmessung), (2009).

[135] V. N. Kuzovkov, E. A. Kotomin, and W. von Niessen. Discrete-lattice theory for Frenkel-defect aggregation in irradiated ionic solids. *Phys. Rev. B*, **58**, 8454, (1998).

[136] U. Natura. Internal report (Schott AG), (2009).

[137] V. M. Orera and E. Alcala. Formation and Size Evolution of Ca Colloids in Additively Colored CaF$_2$. *Phys. Stat. Sol.*, **38**, 621, (1976).

[138] V. M. Orera and E. Alcala. Optical Properties of Cation Colloidal Particles in CaF$_2$ and SrF$_2$. *Phys. Stat. Sol.*, **44**, 717, (1977).

[139] M. L. Sanjuán, P. B. Oliente, and V. M. Orera. The enhaced Raman scattering of phonons in CaF$_2$ and MgO samples containing Ca and Li colloids. *J. Phys. Cond. Matt.*, **6**, 9647, (1994).

[140] R. Bennewitz, C. Günther, M. Reichling, E. Matthias, S. Vijayalakshmi, A. V. Barnes, and N. H. Tolk. Size evolution of low energy electron generated Ca colloids in CaF_2. *Appl. Phys. Lett.*, **66**, 320, (1995).

[141] M. Huisinga, N. Bouchaala, R. Bennewitz, E. A. Kotomin, M. Reichling, V. N. Kurovkov, and W. von Niessen. The kinetics of CaF_2 metallization induced by low-energy electron irradiation. *Nuc. Inst. Meth. Phys. Res. B*, **141**, 79, (1998).

[142] L. P. Cramer, B. E. Schubert, P. S. Petite, S. C. Langford, and J. T. Dickinson. Laser interactions with embedded Ca metal nanoparticles in single crystal CaF_2. *J. Appl. Phys.*, **97**, 074307, (2005).

[143] L. P. Cramer, S. C. Langford, and J. T. Dickinson. The formation of metallic nanoparticles in single crystal CaF_2 under 157 nm excimer laser irradiation. *J. Appl. Phys.*, **99**, 054305, (2006).

[144] C. Wurster, K. Lassmann, and W. Eisenmenger. Resonant phonon scattering by radiation-induced crystal defects in CaF_2. *Phys. Rev. Lett.*, **70**, 3451, (1993).

[145] M. Izerrouken, A. Meftah, and M. Nekkab. Color centers in neutron-irradiated $Y_3Al_5O_{12}$, CaF_2 and LiF single crystals. *J. Lum.*, **127**, 696, (2007).

[146] R. Bennewitz, D. Smith, and M. Reichling. Bulk and surface processes in low-energy-electron-induced decomposition of CaF_2. *Phys. Rev. B*, **59**, 8237, (1999).

[147] T. Tsujibayashi, M. Watanabe, O. Arimoto, M. Itoh, S. Nakanishi, H. Itoh, S. Asaka, and M. Kamada. Two-photon excitation spectra of exciton luminescence in CaF_2 obtained by using synchrotron radiation and laser. *J. Lum.*, **87–89**, 254, (2000).

[148] C. Görling, U. Leinhos, and K. Mann. Surface and bulk absorption in CaF_2 at 193 and 157 nm. *Opt. Commun.*, **249**, 319, (2005).

[149] K. Mann, T. Schröder, B. Schäfer, U. Leinhos, and U. Stamm. Optical pulse stretching and smoothing for ArF and F2 lithography excimer lasers. US Patent 6 389 045, (2002).

[150] K. Kajihara, L. Skuja, M. Hirano, and H. Hosons. Role of Mobile Interstitial Oxygen Atoms in Defects Processes in Oxides: Interconversion between Oxygen-Associated Defects in SiO_2 Glass. *Phys. Rev. Lett.*, **92**, 015504, (2004).

[151] C. F. Bohren and D. R. Huffman. *Absorption and Scattering of Light by Small Particles*, pp. 82–129. John Wiley and Sons, Inc., (1998).

[152] J. R. Tessman, A. H. Kahn, and W. Shockley. Electronic Polarizabilities of Ions in Crystals. *Phys. Rev.*, **92**, 890, (1953).

Die VDM Verlagsservicegesellschaft sucht für wissenschaftliche Verlage abgeschlossene und herausragende

Dissertationen, Habilitationen, Diplomarbeiten, Master Theses, Magisterarbeiten usw.

für die kostenlose Publikation als Fachbuch.

Sie verfügen über eine Arbeit, die hohen inhaltlichen und formalen Ansprüchen genügt, und haben Interesse an einer honorarvergüteten Publikation?

Dann senden Sie bitte erste Informationen über sich und Ihre Arbeit per Email an *info@vdm-vsg.de*.

Sie erhalten kurzfristig unser Feedback!

VDM Verlagsservicegesellschaft mbH
Dudweiler Landstr. 99   Telefon +49 681 3720 174
D - 66123 Saarbrücken   Fax   +49 681 3720 1749
www.vdm-vsg.de

Die VDM Verlagsservicegesellschaft mbH vertritt

Printed by Books on Demand GmbH, Norderstedt / Germany